Report 138                                                                                            1995

# Underground Service Reservoirs: Waterproofing and Repair Manual

R A Johnson
D S Leek
E S King
H B Dunne

CONSTRUCTION INDUSTRY RESEARCH AND INFORMATION ASSOCIATION
6 Storey's Gate London SW1P 3AU
E-mail switchboard@ciria.org.uk
Tel: (0171) 222 8891          Fax: (0171) 222 1708

D
627.86
UND

CIRIA Report 138

# Foreword

The project leading to this Report was carried out, under contract to CIRIA, by Mott MacDonald, Special Services Division, Croydon.

**Research Team**

| | |
|---|---|
| R.A. Johnson CEng MIStructE | Project Director |
| D.S. Leek BSc MSc CEng MIM CGeol FGS MICorr | Project Engineer |
| E.S. King BSc Dipl Eng MPhil CEng MICE | Assistant Project Engineer |
| H.B. Dunne BEng | Materials Engineer |

with contributions from:

| | |
|---|---|
| J.G.M. Wood BSc PhD CEng MICE FIAG | Consultant |
| M.J. Blackler BE PhD | Associate |
| P. Norris BEng MSc PhD CEng MICE | Associate |
| N.W. James BSc CEng MICE MIWEM | Senior Quality Assurance Engineer |
| J. Luckyram MSc DIC PhD | Engineer |
| G. Rajadurai BEng | Engineer |

**Steering Group**

The project was guided and the Manual was prepared with the help of the Project Steering Group.

| | |
|---|---|
| Mr J. Crossley BScTech CEng MICE FIWEM (*Chairman*) | Yorkshire Water. |
| Mr P. Bennison BSc CEng MICE MIStructE FFS | FeRFA & Flexcrete Ltd. |
| Mr R.C. Blacksell CEng MICE MIWEM | Wallace Evans Ltd (for Dŵr Cymru, Welsh Water). |
| Mr R.L. Bonafont BSc CEng MICE | BFRC. |
| Mr K. Cockayne BSc MSc | BASA & Servicised Ltd. |
| Mr P. Edwards BSc CEng MICE ACGI | North West Water. |
| Eur Ing M. Field BSc CEng MICE MIWEM | Southern Projects Ltd. (for Southern Water). |
| Mr P. Homersham MSc MIStructE MIWEM | Thames Water. |
| Mr D. Kemp BSc CEng MICE | Anglian Water. |
| Mr A.J. Parker ARICS | CRA. |
| Mr D.M. Roberts | Wallace Evans Ltd (for Dŵr Cymru, Welsh Water). |
| Mr H. Rosler | BFRC & HT (UK) Ltd. |
| Mr N. Tarbet BSc CEng MPRI | WRc. |
| Mr E. Thorpe CEng MIStructE | South West Water. |

**Corresponding Members**

| | |
|---|---|
| Mr P. Milne | Severn-Trent Water. |
| Mr J.S. Morris | Wessex Water. |
| Mr K. Preston | Northumbrian Water Ltd. |

**CIRIA**

| | |
|---|---|
| Mr J. Sakula | CIRIA Research Manager 1 September 1991 to 23 February 1992. |
| Dr B.W. Staynes | CIRIA Research Manager 24 February 1992 to completion. |

**Financial Support**

The project was financially supported by Anglian Water, Northumbrian Water, North West Water, Severn-Trent Water, South West Water, Southern Water, Thames Water, Welsh Water, Wessex Water and Yorkshire Water.

# Acknowledgements

CIRIA is grateful for information, comments and assistance received from the Steering Group and many others including:

| | |
|---|---|
| Accrete Specialist Contracts Ltd | Mr S. Ballantine |
| Anglian Water | |
| Cemplas Ltd | Mr T. Blenko |
| Concrete Repair Association | |
| Federation of Resin Formulators and Applicators | |
| Fosroc/Expandite | Mr S. Joliffe, Mr I. Moffat, Mr C. Boulton |
| Northumbrian Water | Mr I.D. Moorhouse |
| North West Water | |
| Severn Trent Water | |
| Southern Water Services Ltd | |
| South West Water | Mr D. Horn, Mr C. Low |
| Surveying & Engineering Ltd | Mr S. Humphreys |
| Thames Water Utilities | Mr D. Pratt |
| WRc | |
| Welsh Water | |
| Wessex Water | |
| Yorkshire Water | Mr T. Dyke |
| Mott MacDonald | In-house staff at Birmingham, Cambridge, Glasgow, Manchester and Sheffield |

CIRIA also acknowledges the assistance of over 70 manufacturers and suppliers of waterproofing and repair materials who responded to Mott MacDonald questionnaires.

# Contents

| | |
|---|---|
| Foreword | 1 |
| Acknowledgements | 3 |
| Summary | 4 |
| Contents | 6 |
| List of Figures | 9 |
| List of Tables | 10 |
| | |
| 1 INTRODUCTION | 11 |
|    1.1 Phase I study | 11 |
|    1.2 Phase II study | 11 |
| | |
| 2 STRUCTURE OF THE MANUAL | 12 |
|    2.1 How to use the manual | 12 |
|    2.2 Objectives of the manual | 12 |

## PART A - THE DECISION MAKING PROCESS

| | |
|---|---|
| A1 CONSTRUCTION GROUPS (MATERIALS AND DESIGN) | A1 |
|    A1.1 Roof element construction groups | A4 |
|    A1.2 Wall element construction groups | A5 |
|    A1.3 Column element construction groups | A6 |
|    A1.4 Floor element construction groups | A6 |
| | |
| A2 IDENTIFICATION AND CHARACTERISATION OF DEFECTS | A8 |
|    A2.1 Leakage through roofs | A8 |
|    A2.2 Leakage through walls and floors | A8 |
|    A2.3 Reasons for investigation | A9 |
|    A2.4 Defect categories, causes and effects | A10 |
| | |
| A3 METHODS OF REPAIR AND MATERIAL CATEGORIES | A15 |
|    A3.1 Corrosion of structural metalwork | A16 |
|    A3.2 Concrete and brickwork deterioration | A17 |
|    A3.3 Cracking (other than formed joints) | A19 |
|    A3.4 Corrosion of reinforcement | A21 |
|    A3.5 Joint failure | A22 |
|    A3.6 Breakdown of waterproofing systems | A24 |
|    A3.7 Problems with access (including access ladders) ventilation, openings and ancillary items | A25 |
| | |
| A4 REPAIR MATERIALS | A30 |
|    A4.1 Materials approval schemes | A30 |
|    A4.2 Material categories | A32 |
|    A4.3 Materials selection | A51 |

| APPENDIX AP I CAUSES OF DEFECTS IN RESERVOIRS | A52 |
|---|---|
| AP I.1 Corrosion of metalwork (including structural metalwork and access ladders, etc.) | A52 |
| AP I.2 Concrete and brickwork deterioration | A54 |
| AP I.3 Cracking (other than formed joints) | A59 |
| AP I.4 Corrosion of reinforcement | A62 |
| AP I.5 Breakdown of waterproofing systems | A62 |
| APPENDIX AP II VENTILATION REQUIREMENTS FOR RESERVOIRS | A64 |
| AP II.1 Derivation of the ventilation provision formula | A64 |

# PART B - CONSIDERATIONS IN THE DECISION MAKING PROCESS

| | |
|---|---|
| B1 SAFETY OF OPERATIONS | B1 |
| B1.1 Particular hazards applicable to working in and around service reservoirs | B1 |
| B1.2 Responsibilities for safety | B1 |
| B1.3 Planning, personnel selection and training | B2 |
| B1.4 Confined space working and permits to work | B3 |
| B2 DRAINAGE PROVISIONS | B5 |
| B2.1 Drainage provisions on roofs | B5 |
| B2.2 Drainage provisions for walls | B6 |
| B2.3 Drainage provisions for floors | B7 |
| B3 AVOIDANCE OF DEFECTS ARISING FROM REPAIR ACTIVITY | B8 |
| B3.1 Structural problems arising through repair activity | B8 |
| B3.2 Problems associated with repair methods and materials | B12 |
| B4 ROOF INSULATION AND EARTHWORKS | B17 |
| B4.1 Roof insulation | B17 |
| B4.2 Earthworks | B18 |
| B5 DURABILITY, MAINTENANCE AND COST OF REPAIRS AND WATERPROOFING | B19 |
| B5.1 Durability of repairs and waterproofing | B19 |
| B5.2 Maintenance | B20 |
| B5.3 Cost | B21 |
| B6 EXAMPLE MATERIAL PERFORMANCE REQUIREMENTS | B23 |
| B6.1 Concrete repair system | B24 |
| B6.2 Waterproofing membrane on the outside | B25 |
| B6.3 Internal waterproofing treatments | B29 |
| B6.4 Surface protective coatings | B30 |
| B6.5 Metalwork paint systems | B31 |
| B6.6 Crack injection | B32 |
| B6.7 Crack sealants | B33 |
| B6.8 Joint sealants | B34 |
| B6.9 Large volume grouting | B34 |

# PART C - INFORMATION TO AID THE DECISION MAKING PROCESS

| | |
|---|---|
| C1 CONTAMINATION OF RESERVOIRS | C1 |
| C1.1 Legal requirements | C1 |
| C1.2 Water treatment | C1 |
| C1.3 Contamination of treated water | C2 |

| | |
|---|---|
| **C2 INVESTIGATION AND TESTING METHODS** | C5 |
| C2.1 Review of the existing plans | C6 |
| C2.2 Visual survey | C6 |
| C2.3 Detailed investigation | C8 |
| **C3 CONTRACT DOCUMENTATION** | C14 |
| C3.1 Preprinted documentation | C14 |
| C3.2 Suitability of documentation for waterproofing and repairing service reservoirs | C16 |
| C3.3 Contract format | C17 |
| **C4 QUALITY MANAGEMENT IN REPAIRS TO SERVICE RESERVOIRS** | C18 |
| C4.1 Introduction | C18 |
| C4.2 During the investigation and condition assessment | C18 |
| C4.3 During design of the repair | C19 |
| C4.4 During materials selection and testing | C19 |
| C4.5 During the repair | C20 |
| C4.6 Quality scheme for repair contractors | C22 |
| **C5 GUIDANCE FOR FUTURE DESIGN** | C23 |
| C5.1 Drainage | C23 |
| C5.2 Protection of structural metalwork | C23 |
| C5.3 Avoidance of concrete deterioration | C23 |
| C5.4 Avoidance of cracking (other than formed joints) | C24 |
| C5.5 Avoidance of corrosion of reinforcement | C24 |
| C5.6 Avoidance of joint failure | C25 |
| C5.7 Durability of waterproofing membranes | C25 |
| C5.8 Access (including access ladders), ventilation, openings and ancillary items | C25 |
| C5.9 Waterstops | C26 |
| C5.10 Durability of sealants | C26 |
| **REFERENCES** | C27 |
| **INDEX** | C31 |

## List of Figures

|  |  | Page |
|---|---|---|
| 2.1 | Steps in the decision making process | 15 |
| A1.1 | Construction groups for reservoirs - chronological reference | A3 |
| A3.1 | Sealing of non-moving cracks up to 1 mm wide | A20 |
| A3.2 | Methods of sealing moving cracks | A20 |
| A3.3 | Pond testing a small area on a reservoir roof | A25 |
| A3.4 | Vandal resistant access design | A26 |
| A3.5 | Vandal resistant combined access and vent | A27 |
| A3.6 | Typical details for vandal-proof vents | A27 |
| AP I.1 | Effect of carbonate imbalance on concrete within a reservoir. | A58 |
| AP I.2 | Efflorescence on reservoir roof and walls | A59 |
| AP I.3 | Root penetration into a reservoir | A61 |
| AP II.1 | Reservoir ventilation - schematic diagram of a reservoir drained by gravity | A64 |
| B1.1 | Typical sequence of operations for working in a confined space | B4 |
| B2.1 | Typical roof edge drainage details | B6 |
| B3.1 | Cracks at the centre span of vaulted roofs | B9 |
| B3.2 | Expansion of roof due to heating | B9 |
| B3.3 | Arching of roof due to heating | B10 |
| B3.4 | Upward flexing of untied barrel arches | B10 |
| B3.5 | A section of a membrane has been removed at a joint showing adhesion failure between a membrane and substrate due to inadequate surface preparation, which led to leakage | B13 |
| B3.6 | Rippling in membranes | B14 |
| B3.7 | Recracking of a 'sealed' crack | B16 |
| C2.1 | Stages in the investigation of a service reservoir | C5 |
| C2.2 | Measurement of the depth of carbonation of concrete | C9 |
| C4.1 | Outline flow diagram for design stage activities (for each site or part-site) | C19 |
| C4.2 | Outline flow diagram for materials selection and testing activities | C20 |
| C4.3 | Outline flow diagram for installation stage activities - Contractor | C21 |
| C4.4 | Outline flow diagram for installation stage activities - Employer | C21 |

# List of Tables

|  |  | Page |
|---|---|---|
| 2.1 | Repair and waterproofing methodology. | 12 |
| 2.2 | Relationship between the three parts of the Manual | 13 |
| 2.3 | Steps in the decision making process | 16 |
| 2.4 | Considerations in the decision making process | 17 |
| 2.5 | Information to aid the decision making process | 18 |
| A1.1 | Principal construction type as percentage of existing service reservoirs. | A1 |
| A2.1 | Defect categories identified | A10 |
| A4.1 | Relationship between rate of cure and movement accommodation factor for different generic groups of sealants | A48 |
| AP I.1 | Galvanic series of metals and alloys (abbreviated) in a strong electrolyte. | A54 |
| B5.1 | Indicative estimates of the durability of materials | B20 |
| B5.2 | Typical applied unit costs of waterproofing and repair materials | B22 |

# 1 Introduction

## 1.1 PHASE I STUDY

Technical Note 145 reported on the first phase of the two-stage study on waterproofing and repairing underground service reservoirs. Its objective was to produce an authoritative guide to assist Water Undertakers dealing with reservoir roofs constructed of brickwork, mass or reinforced concrete, or a combination of these materials, in need of repair and/or waterproofing. The Technical Note included guidance for dealing with problems of leakage, deterioration of materials and corrosion of metals in the roof, unsubmerged walls and the roof/wall joint.

Practical and authoritative guidance with clear identification of areas of uncertainty was presented on all relevant aspects of waterproofing and concrete repair.

## 1.2 PHASE II STUDY

It was recognised during the Phase I study that operators of service reservoirs were experiencing problems related to all elements of the structure.

The scope of the Phase I study was, however, restricted to problems of leakage through the roof, the upper sections of walls (above water level) and the roof/wall joint. The final published report (TN145) identified a number of areas where the scope of the study could usefully be extended, to include other sections of the reservoir (where no guidance was currently available) and to monitor feedback from the users of the Technical Note (to provide supplementary advice where necessary).

This report, written in guidance manual format, is the result of feedback from the dissemination of the Phase 1 report within the water utilities and repair industry. Its objective is to assist Water Undertakers to carry out effective investigations, repairs and waterproofing with improved quality and increased cost-effectiveness.

# 2 Structure of the manual

## 2.1 HOW TO USE THE MANUAL

This manual has been written in three parts:

Part A        Steps in the decision making process
Part B        Considerations in the decision making process
Part C        Information to aid the decision making process.

To aid use of the manual, each Part begins with a coloured sheet and has separate page numbering: i.e. Part A starts with page A1.

> Text that appears in shaded boxes is supplementary to the main text and is either advice or additional information.

## 2.2 OBJECTIVES OF THE MANUAL

Information in this manual has been divided into three Parts, the objectives of which are outlined in Table 2.1. Each Part is subdivided into Sections.

**Table 2.1** Repair and waterproofing methodology

| Part | Objective |
|---|---|
| **PART A**<br><br>STEPS IN THE DECISION MAKING PROCESS | To provide the technical information essential in the decision making process, for the repair and waterproofing of service reservoirs, by following the steps below:<br><br>STEP 1 Identify the **construction group**<br>STEP 2 Identify the **defect category**<br>STEP 3 Identify the most suitable **method of repair** and **material category**<br>STEP 4 Select repair material<br><br>Section 2.2.1 gives the objective of each step and a description of the advice furnished by the manual. |
| **PART B**<br><br>CONSIDERATIONS IN THE DECISION MAKING PROCESS | To highlight other important considerations in repair and waterproofing work which, although not necessary for the decision making process, are essential reading for anyone using this manual.<br><br>Further detail is given in Section 2.2.2. |
| **PART C**<br><br>INFORMATION TO AID THE DECISION MAKING PROCESS | To provide the reader with information that will aid the decision making process.<br><br>Further detail is given in Section 2.2.3 |

Table 2.2 shows the stages at which reference should be made to each Section within each Part of the manual.

**Table 2.2** Relationship between the three parts of the Manual

| | | PART A STEPS IN THE DECISION MAKING PROCESS | | | |
|---|---|---|---|---|---|
| | | STEP 1 | STEP 2 | STEP 3 | STEP 4 |
| | | Identify construction group | Identify defect category | Identify most suitable method of repair and **material category** | Select repair material |
| | Section | A1 | A2 | A3 | A4 |
| **PART B** <br> **CONSIDERATIONS IN THE DECISION MAKING PROCESS** | | | | | |
| Safety of operations | B1 | ■ | ■ | ■ | ■ |
| Drainage provisions | B2 | ■ | ■ | ■ | |
| Defects arising from repair activity | B3 | | | ■ | |
| Roof insulation and earthworks | B4 | | ■ | ■ | |
| Durability, maintenance and costs | B5 | | | ■ | ■ |
| Example material performance requirements | B6 | | | | ■ |
| **PART C** <br> **INFORMATION TO AID THE DECISION MAKING PROCESS** | | | | | |
| Contamination of reservoirs | C1 | | ■ | ■ | ■ |
| Investigation and testing methods | C2 | | ■ | | |
| Contract documentation | C3 | | | | |
| Quality management | C4 | | Relevant to all steps | | |
| Guidance for future design | C5 | | | | |

### 2.2.1 Steps in the decision making process

The objective and contents in the manual for each of the Steps within Part A are summarised in Table 2.3. Details of the **construction groups, defect categories, methods of repair** and **materials categories** are given in Figure 2.1.

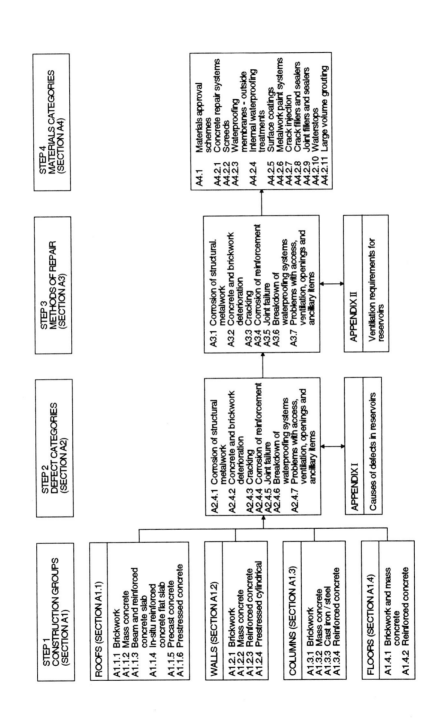

**Figure 2.1** *Steps in the decision making process*

**Table 2.3** Steps in the decision making process

| Steps | Section reference | Objective | Manual content |
|---|---|---|---|
| STEP 1 Identify **construction group** | A1 | To identify reservoir construction types to enable correct identification of defects and selection of suitable methods for repair and materials. | Each of the construction groups listed in Figure 2.1 is described giving information relevant to waterproofing and repair. |
| STEP 2 Identify **defect category** | A2 | To assist in defect identification so that appropriate methods of repair and materials may be selected. | For each of the defect categories listed in Figure 2.1, the construction groups to which it might apply, the effects of the particular defect, and their causes are listed (it refers to Appendix I, which details the mechanisms of deterioration). |
| STEP 3 Identify most suitable **method of repair** and **material category** | A3 | To enable the most effective method of repair to be selected for a particular construction group and defect. | For each defect, different methods of repair are discussed with respect to suitable applications. Relevant materials categories, from the list in Figure 2.1, are identified. |
| STEP 4 Select repair material | A4 | To enable the most effective material to be selected for a particular construction group and defect category. | Materials within each material category are discussed in generic terms. Advice is given as to their suitability, use and specification. |
| Appendix I | AP I.1 to AP I.5 | To enable the mechanisms of deterioration within each defect category to be identified. | The mechanisms of deterioration within each defect category are described. |
| Appendix II | AP II.1 | To show the derivation of the formula for reservoir ventilation. | The derivation of the formula for reservoir ventilation requirements is derived from first principles. |

### 2.2.2 Considerations in the decision making process

The objective and contents in the manual for each of the Considerations within Part B are summarised in Table 2.4.

**Table 2.4** Considerations in the decision making process

| Consideration | Section reference | Objective | Manual content |
|---|---|---|---|
| Safety of operations | B1 | To make the reader aware of the safety aspects of undertaking repair and waterproofing work to a service reservoir. | Good practice in working both on the roof and inside service reservoirs, and on handling and application of materials on site. |
| Drainage provisions | B2 | To make the reader aware of the importance of good drainage. | Defects that arise in drainage, corrective action and methods for providing new drainage systems. |
| Avoidance of defects arising from repair activity | B3 | To enable the reader to appreciate possible problems that could arise through repair activity and with the materials used. | Examples of potential problems and advice on their avoidance. |
| Roof insulation and earthworks | B4 | To describe alternative forms of roof insulation and earthworks. | The advantages and disadvantages of various types of roof insulation and roof uses are presented. |
| Durability, maintenance and cost of repairs and waterproofing | B5 | To provide a basis for long-term planning of reservoir maintenance and repair programmes. | Expected service lives of materials, recommended maintenance procedures and unit costs of waterproofing and repair materials. |
| Example material performance requirements | B6 | To provide example performance requirements which could be developed into particular specification clauses for repair and waterproofing materials. | Example performance requirements (in specification clause format). |

### 2.2.3 Information to aid the decision making process

The objective and contents in the manual for each of the Information Sections within Part C are summarised in Table 2.5.

**Table 2.5** Information to aid the decision making process

| Information | Section reference | Objective | Manual content |
|---|---|---|---|
| Contamination of reservoirs | C1 | To provide a background on the requirements for potable water. | Legal requirements and treatment of potable water and how it may become contaminated. |
| Investigation and testing methods | C2 | To assist in identifying a defect, its cause and its extent. | Details of inspection, investigation and testing techniques. |
| Contract documentation | C3 | To offer advice on forms of contract and on contract documentation. | A description of appropriate forms of contract and a check list of what to include in contract documentation. |
| Quality management in repairs to service reservoirs | C4 | To introduce the reader to quality management systems appropriate to reservoir waterproofing and repair. | Quality management systems for investigation and condition assessment, design, materials selection and testing, repair work and repair contractors. |
| Guidance for future design | C5 | To offer advice for future design and construction of service reservoirs. | A list of advice points. |

# PART A

# The decision making process

SECTION A1 CONSTRUCTION GROUPS (MATERIALS AND DESIGN)
SECTION A2 IDENTIFICATION AND CHARACTERISATION OF DEFECTS
SECTION A3 METHODS OF REPAIR AND MATERIALS CATEGORIES
SECTION A4 REPAIR MATERIALS

APPENDIX I CAUSES OF DEFECTS IN RESERVOIRS
APPENDIX II VENTILATION REQUIREMENTS IN RESERVOIRS

# A1 Construction groups (materials and design)

> Considerations in the decision making process that should be read in conjunction with this Section are: Sections B1 and B2.
>
> Information aiding the decision making process relevant to this Section can be found in: Sections C3, C4 and C5.

The earliest covered reservoirs were constructed in the middle of the nineteenth century as a result of growing concern over public health and the quality of water stored in open reservoirs. Many of these early reservoirs were of brick construction, but materials and construction techniques have changed over the years to follow developments in technology and contemporary economic considerations.

A survey reported by the WRc[2] in 1988 gave the relative percentages of each reservoir construction group. A summary of the results is given in Table A1.1.

**Table A1.1** Principal construction type as percentage of existing service reservoirs

| Construction group | Proportion identified in the survey (approx.) |
| --- | --- |
| Brickwork | 19.0% |
| Mass concrete | 21.0% |
| Reinforced concrete | 57.0% |
| Post-tensioned concrete | 2.0% |
| Others | 1.0% |
| Glass-reinforced plastic (roof only) | Known to exist but not included in the survey |
| Aluminium (roof only) | Known to exist but not included in the survey |

As the table shows, there were a small number of reservoirs with lightweight roofs constructed from glass-reinforced plastic or aluminium. Historically, many open reservoirs had roofs 'added' from these materials in the period from the mid 1930s to the mid 1950s (the practice in some areas continued until the early 1970s), often with an unsatisfactory 'fit'. Timber and coated steel roofs are also known to exist. Such construction groups are rare, and as they are unlikely to be covered, they are considered to be outside of the scope of this document, although the general advice given in a number of sections will be of use when undertaking repair or replacement works. There are also other 'special case' constructions, particularly for roof construction groups e.g. *in-situ* reinforced concrete domes, with post-tensioned perimeters and unbonded tendons, which it has not been possible to fully encompass.

Consequently, there are a large number of unique reservoir designs for which it is not possible to identify individual problems and solutions. However, broad construction groups have been identified for the four principal construction elements, roof, walls, columns and floor, as follows.

**Roof element construction groups**

1. brickwork
    (a) barrel vault
    (b) domed
2. mass concrete
3. beam and reinforced concrete slab
    (a) filler joist
    (b) steel joist beams
    (c) reinforced concrete beams
4. in-situ reinforced concrete flat slab
5. precast concrete
6. prestressed concrete.

**Wall element construction groups**

1. brickwork
2. mass concrete
    (a) with brick lining
    (b) gravity (resultant pressure in the middle ⅓)
    (c) reinforced
3. reinforced concrete
    (a) cantilever
    (b) propped cantilever
    (c) spanning vertically or horizontally
    (d) double walled
4. prestressed cylindrical.

**Column element construction groups**

1. brickwork
2. mass concrete
3. cast iron/steel
4. reinforced concrete.

**Floor construction groups**

1. brickwork and mass concrete
2. reinforced concrete
    (a) single course
    (b) two course.

Figure A1.1 shows the period over which each element construction group was generally built and can be used as an initial guide if the form of construction is unknown. Construction groups should then be confirmed by inspection or investigation.

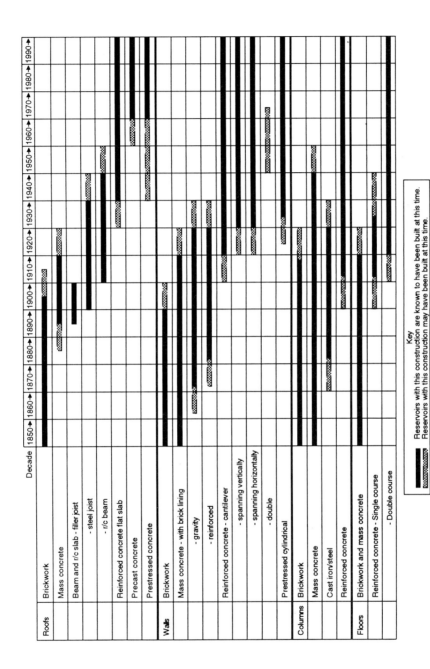

**Figure A1.1** Construction groups for reservoirs - chronological reference

CIRIA Report 138

A3

## A1.1 ROOF ELEMENT CONSTRUCTION GROUPS

### A1.1.1 Brickwork

Most brickwork roofs were constructed in the period between 1850 and 1905. They were usually constructed as parallel-sided barrel arches, segmental-shaped arches in circular structures or domes.

Roof support within the reservoir was by one of the following means:

- Brickwork walls with arch openings.
- Wrought or cast iron joists or girders supported on cast iron, wrought iron or brickwork columns. In some cases the girders were tied together with wrought iron rods.

The spandrels between arches were usually filled with concrete, the quality of which varied. The concrete infill was sometimes designed to be structural, to strengthen the roof, and would therefore be of reasonable quality; at the other extreme the infill concrete was often very lean and permeable.

In some cases waterproofing was laid on the roof as part of the original construction. This took the form of either a lime render, a tar coating or an asphalt layer about 1 inch (25 mm) thick. In some instances (and as recently as 1956), a layer of puddled clay was laid on the roof, a typical thickness being 9 inches (225 mm); this was especially common in London.

### A1.1.2 Mass concrete

Some open reservoirs, constructed in the nineteenth and early twentieth centuries, were later covered by flat or arched mass concrete roofs. In such cases, the reason for selecting this material instead of brickwork was often to reuse the aggregate from obsolete filter beds. The constructions used for roof support were similar to those employed for brickwork roofs. An external cementitious render was commonly used for waterproofing.

### A1.1.3 Beam and reinforced concrete slab

Reinforced concrete came into use for reservoir roof construction at the turn of the century. The earliest types of reinforcement used were steel fabric or expanded mesh. The construction was generally flat, or shallow, arched reinforced slabs, spanning between rolled steel joists which were concrete encased (except for the bottom flange). Another form of construction was filler joist, with the joists closely spaced (400–600 mm) and built into the concrete roof slab, the latter being either plain or reinforced with steel mesh. The use of reinforcing steel bars, introduced *circa* 1910, increased more rapidly in slabs than in the supporting beams. Roof slabs were generally supported by steel beams until the 1940s, when a steel shortage and rising costs increased the use of reinforced concrete beams.

One exception to this was a design, popular between 1900 and 1930, that consisted of hexagonal mass concrete domes supported on reinforced concrete groynes, which were connected at the columns by milled steel plates. These roofs were drained by asphalt lined channels in the springing.

### A1.1.4 *In-situ* reinforced concrete flat slab

Once the progression had been made to the use of reinforced concrete beams, it was a natural step to start designing roofs as flat slabs. This gave the advantages of easier (and thus cheaper) soffit shuttering during construction and a reduction in the number of necessary air vents due to the elimination of downstand beams.

Flat concrete roofs were sometimes waterproofed with a layer of asphalt, rubber bitumen emulsion, incorporating glass fibre or a bonded membrane. Rendering with cementitious mortar to improve the surface finish of the concrete provided some degree of waterproofing.

### A1.1.5 Precast concrete

There have been several permutations of *in-situ*/precast element roof designs. Two of the most common are:

- in-situ beams between columns supporting precast slabs
- precast beams between columns supporting *in-situ* slabs or precast slabs joined with strips of *in-situ* concrete.

Waterproofing, which in the past usually took the form of an asphalt layer, was often considered to be essential on roofs incorporating precast elements because of the large number of joints.

### A1.1.6 Prestressed concrete

The use of prestressing in reservoir roofs started in the late 1960s. Typical constructions were:

- prestressed beams between columns with *in-situ* slabs
- prestressed beams and precast reinforced concrete slabs
- post-tensioned flat slab roofs
- *in-situ* reinforced concrete dome, with post-tensioned perimeter.

Asphalt waterproofing was also often employed on these structures.

## A1.2 WALL ELEMENT CONSTRUCTION GROUPS

### A1.2.1 Brickwork

The use of brickwork for walls (as with roofs) was prominent in the latter half of the nineteenth century. They were designed as gravity walls, horizontal arches or a combination of both. Waterproofing was usually on the outside and generally consisted of a layer of puddled clay, which was kept saturated by allowing a small amount of the contained water to weep through the walls, by leaving a pattern of open perpends.

Brickwork walls at this time were usually designed with inadequate factors of safety and most relied on the passive resistance of earth to withstand water pressure and the thrust from arched roofs.

### A1.2.2 Mass concrete

Mass concrete wall construction was also prominent in the latter half of the nineteenth century, but because of its greater versatility and ability to be waterproofed its use outlasted brickwork construction by up to 30 years. Many of these reservoirs were waterproofed by a cementitious render or an asphalt lining.

The early designs were often lined internally with brickwork, either for aesthetic reasons or to cover an asphalt waterproofing layer. It was also common to provide an external puddled clay waterproofing layer.

Circular mass concrete tanks, which relied on the passive resistance of the earth to resist internal water pressure, were also constructed.

### A1.2.3 Reinforced concrete

*Cantilever*

The cantilever wall generally takes the form of a vertical slab diminishing in thickness from base to top. The base slab is usually extended as far as possible into the reservoir and the overall shape is either that of an inverted T or an L. The structure is designed so that the resultant pressure lies within the middle third of the base. The roof of the reservoir is free to move relative to the walls.

*Spanning vertically*
A vertically spanning wall is designed as though it is a slab supported by the roof and the base slab. Shear connection is necessary at the roof/wall joint. A special case is the propped cantilever, where the roof support is pinned: i.e. it offers no resistance to moments.

*Spanning horizontally*
A horizontally spanning, or counterfort, wall is essentially a continuous horizontal or inclined slab supported by vertical counterforts. The wall is cantilevered vertically but can be designed for a greater height than a simple cantilever design.

*Double walled*
A double wall prevents reverse pressure on the inner wall and leakage of groundwater into the reservoir. The space inside the wall can be used for drainage, inspection and repair.

### A1.2.4 Prestressed cylindrical

For small reservoirs a cylindrical wall was often economical. The two most common methods of construction were:

- Lightly reinforced sprayed concrete walls cast against a shutter with wire wound around the outside, which was stressed during winding. The wires were then covered by a second layer of sprayed concrete.

- *In-situ* concrete post-tensioned walls.

## A1.3 COLUMN ELEMENT CONSTRUCTION GROUPS

### A1.3.1 Brickwork
Brickwork columns outlived the use of this material for roofs and walls and continued into the 1920s.

### A1.3.2 Mass concrete
Mass concrete columns constructed either as precast blocks or as a single pour were superseded by reinforced concrete.

### A1.3.3 Cast iron/steel
The use of cast iron for roof support columns was prevalent between 1880 and 1910, but was replaced by reinforced concrete early in the twentieth century.

Some columns, pre-1940, were constructed from rolled steel, protected from corrosion by a bituminous coating.

### A1.3.4 Reinforced concrete
Reinforced concrete has been used almost exclusively for columns since approximately 1920.

## A1.4 FLOOR ELEMENT CONSTRUCTION GROUPS

### A1.4.1 Brickwork and mass concrete
The earliest reservoirs were paved in brickwork, quarry tiles or stone flagging and were externally lined with a layer of puddle clay. Later it became accepted practice to lay a concrete blinding layer to provide a clean even surface on which to lay the floor proper.

During the late nineteenth century floors were either brickwork over a layer of mass concrete, or solid mass concrete. Both groups disappeared early in the twentieth century once reinforced concrete came into general use.

Some early concrete floors were provided with an asphalt lining for additional waterproofing.

### A1.4.2 Reinforced concrete

Initially reinforced concrete floors were laid in one course, sometimes with steel joists inserted under and between columns to carry higher loads. Where floors were founded on clean rock it was found that a two-course floor became desirable, as the restraint of the rock would result in the development of cracks due to shrinkage. A structural two-course floor was also the natural progression from the early use of blinding layers. True two-course floor construction has been adopted in most reservoirs constructed since 1920.

There are three different types of reinforced concrete floor:

- Square block: Essentially the floor is cast in bays between column footings and the joints between bays sealed.
- Reinforced slab: Required where foundation or uplift problems are likely to occur. Properly designed movement joints must be provided.
- Groyned arches: The arches transmit hydrostatic pressure to the columns.

# A2 Identification and characterisation of defects

> Considerations in the decision making process which should be read in conjunction with this Section are: Sections B1, B2 and B4.
>
> Information aiding the decision making process relevant to this Section can be found in: all Sections of Part C.

## A2.1 LEAKAGE THROUGH ROOFS

A major problem associated with underground raw and potable water storage reservoirs is leakage through the roof or the upper parts of the walls, particularly at joints, resulting in contamination of the water supply: see Section C1. In many cases, the ingress of groundwater has also led to structural deterioration, e.g. corrosion of reinforcement and spalling of soffit concrete. In other cases an existing structural problem, e.g. cracks caused by changes in the stress state of the roof, has led to leakage.

Leakage into a reservoir can be detected by failed water quality samples, or regular inspections, which may identify groundwater ingress (potential water quality problems) before an unsatisfactory analysis has been obtained. The investigation and testing triggered by these tests and the remedial measures often required to return the reservoir to an acceptable condition are described in the following sections.

## A2.2 LEAKAGE THROUGH WALLS AND FLOORS

When a service reservoir is operational and is full, ingress of groundwater is generally restricted to the upper sections of the walls, particularly the roof/wall joint. Internal hydrostatic pressure usually ensures that leakage through the lower sections of the wall and/or floor (again predominantly through joints) is outward from the reservoir. Ingress can, and does, occur when the reservoir is not full (the water level may vary by 40% or more) or when the reservoir has been drained down for inspection or remedial works. There are examples where reservoirs have been built above, or just downstream of natural springs, where the external hydrostatic pressure is sufficiently great to overcome the internal pressure and net movement of water into the reservoir has occurred.

The consequences of groundwater ingress through the walls and floor are identical to those for ingress through the roof: i.e. failure to meet the water quality regulations, resulting in the requirement for remedial work to rectify the situation. Leakage out through the walls and floor may be high (up to 1 Ml per day in a 10 Ml reservoir if operating full) but does not compromise the statutory regulations; hence repairs to correct this situation are not mandatory. Repairs are usually carried out when the outward leakage becomes unmanageable or uneconomic: i.e. when repair is cost effective.

## A2.3 REASONS FOR INVESTIGATION

Regular inspection of reservoirs to assess structural or durability performance is not routine for some Water Undertakers. No statutory requirement exists for inspection of the majority of reservoirs currently in service. Routine monitoring of water quality may be considered adequate. Instances of inspection to detect leakage into or out of a reservoir include the following.

1. Routine inspection under the Reservoirs Act 1975[3].
2. 'In-house' inspection of reservoirs not covered by the Act.
3. When unsatisfactory bacteriological counts are obtained from routine water quality tests (see Section C1.1).
4. When variations in a number of other water quality parameters exist, e.g. taste or odour changes, physico-chemical or other undesirable parameters (see Section C1.1).
5. When the extent of water loss becomes unacceptable.

Under the 1975 Reservoirs Act, the owners of reservoirs with a *'capacity greater than 25,000 cubic metres of water above the natural level of any part of the land adjoining the reservoir'* have a legal obligation to inspect their structures at a specified interval, which must not exceed ten years. This inspection is carried out by a statutory Panel Engineer appointed by the Secretary of State for the Environment. The traditional sources of information available to the inspecting engineer include previous visual inspection reports and a review of the historic records of the structure. Subject to a satisfactory inspection, the reservoir is granted an authorised life of a further ten years; however, the inspecting engineer is empowered to order remedial works if they are deemed to be necessary. These inspections are for structural purposes and will not specifically identify defects that are likely to affect water quality.

Reservoirs below this size are exempt from inspection under the provisions of the Act and no statutory obligation for regular inspections exists. Some Water Undertakers, however, have instigated programmes of routine 'in-house' inspections, by company staff or specialist contractors, in order to identify defects before they become detrimental. The frequency of these 'in-house' inspections varies greatly between the different Water Undertakers.

In many cases, particularly for buried reservoirs, where external inspection of the structure is not possible, the first indication of a structural problem arises from unsatisfactory bacteriological counts obtained during routine testing of the water. The presence of coliform bacteria in the water are usually indicative of the breakdown of the integrity of the reservoir and contamination of its contents. Breakdown often occurs rapidly and prediction from routine inspections may not have been possible.

> Contamination requires immediate investigation, specification and remedial work in order that the statutory requirements for water quality can be met.

Changes in other water quality parameters may also instigate investigation. In some cases, investigation may show that the contamination had occurred as a result of deterioration of construction or repair materials, e.g. lead (which was used in caulking around manholes), red lead coatings, timber depth gauges, etc., rather than from loss of integrity of the structure.

> Where an investigation identifies materials used in the original construction, or in previous repairs, known to result in water quality problems, they should be removed and replaced.

Outward leakage from a reservoir is generally regarded as a lower priority than water quality when undertaking an inspection, unless the extent of water loss has become excessive. In many cases it is difficult to identify whether the leakage is occurring from the reservoir or its associated pipework. Water loss may also be due to overflow resulting from inoperative ball valves.

## A2.4 DEFECT CATEGORIES, CAUSES AND EFFECTS

The objective of this section is to help Water Undertakers to identify a defect, to understand the cause of that defect, and to categorise the defect for purposes of selecting the most appropriate repair strategy.

Construction groups that have been identified are detailed in Section A1.

In many situations, the category, location and cause of a defect will be evident from an internal inspection of the reservoir. However, in some cases (e.g. cracks in brickwork, outward leakage) defects can be difficult to detect. This section briefly describes the most common defects that have been found in reservoirs and gives an indication of the defect categories associated with each construction group. Unless a defect can be accurately located and its cause and full effect understood, it will not be possible to select the most effective method of repair.

In Sections A2.4.1 ~ A2.4.7 seven **defect categories** have been identified together with their most common **causes** and **effects**. Detailed information on the mechanisms of deterioration associated with some of the defect categories identified is described in Appendix I (located at the end of Part A).

**Table A2.1** Defect categories identified

| Defect categories (including Section reference) | Appropriate section in Appendix I |
| --- | --- |
| A2.4.1 Corrosion of structural metalwork | AP I.1 Corrosion of metalwork (including structural metalwork and access ladders, etc.) |
| A2.4.2 Concrete and brickwork deterioration | AP I.2 Concrete and brickwork deterioration |
| A2.4.3 Cracking (other than formed joints) | AP I.3 Cracking (other than formed joints) |
| A2.4.4 Corrosion of reinforcement | AP I.4 Corrosion of reinforcement |
| A2.4.5 Joint failure | No equivalent |
| A2.4.6 Breakdown of waterproofing systems | AP I.5 Breakdown of waterproofing systems |
| A2.4.7 Problems with access (including access ladders), ventilation, openings and ancillary items. | No equivalent; see also AP I.1 |

### A2.4.1 Corrosion of structural metalwork

*Relevant construction groups*

| Roof | Wall | Column | Floor |
|---|---|---|---|
| Brickwork<br>Mass concrete<br>Beam and reinforced concrete slab | Mass concrete with embedded steel sections | Cast iron<br>Rolled steel | None |

*Causes*
- exposure to the moist environment above the water level within the reservoir
- exposure to stored water within the reservoir
- bimetallic corrosion
- exposed metalwork above roof level.

> Details of the types and mechanisms of corrosion are given in Appendix I, Section AP I.1.

*Effects*
- loss of structural integrity
- possible contamination of the water by loose corrosion products, or paint
- medium for bacterial growth
- safety hazard.

### A2.4.2 Concrete and brickwork deterioration

*Relevant construction groups*

| Roof | Wall | Column | Floor |
|---|---|---|---|
| All | All | Brickwork<br>Mass concrete<br>Reinforced concrete | All |

*Causes*
- poor workmanship
- chemical attack
- physical attack.

> Details of the individual causes are given in Appendix I, Section AP I.2.

*Effects*
- loss of protection to reinforcement
- seepage of water in or out of the reservoir
- loss of structural integrity
- softening of concrete
- spalling of concrete
- loss of mortar from joints in brickwork
- detachment of brickwork linings from mass concrete walls.

## A2.4.3 Cracking (other than formed joints)

*Relevant construction groups*

| Roof | Wall | Column | Floor |
|------|------|--------|-------|
| All | All | Brickwork<br>Mass concrete<br>Reinforced concrete | All |

*Causes*
- plastic and drying shrinkage
- plastic settlement
- changes in loading and/or stress state
- sulphate attack
- reactive aggregate
- salt crystallisation
- freeze-thaw
- corrosion of reinforcement
- removal of supporting ground
- ground settlement/subsidence
- thermal movements
- high alumina cement.

> For details see Appendix I, Section AP I.3.

*Effects*
- leakage inwards or outwards from the reservoir
- loss of structural strength
- excessive deflection or excessive joint opening or closing.

## A2.4.4 Corrosion of reinforcement

*Relevant construction groups*

| Roof | Wall | Column | Floor |
|------|------|--------|-------|
| Beam and reinforced concrete slab<br>*In-situ* reinforced concrete flat slab<br>Precast concrete<br>Prestressed concrete | Reinforced concrete<br>Prestressed cylindrical<br>Post-tensioned | Reinforced concrete | Reinforced concrete |

*Causes*
- chemical attack, resulting in loss of alkalinity of the concrete from
  - carbonation
  - acid attack
  - carbonate imbalance
- chloride attack
- cracking
- exposure of prestressing bands to water leaking outward or groundwater.

> For details see Appendix I, Section AP I.4.

*Effects*
- loss of structural integrity due to loss of steel section
- loss of bond and reduction in concrete cross-section as a result of spalling
- cracking of concrete due to expansive corrosion products.

### A2.4.5 Joint failure

*Relevant construction groups*

| Roof | Wall | Column | Floor |
|---|---|---|---|
| All | All | None | All |

*Causes*
- poor workmanship and detailing including
  - incorrect positioning of the joint
  - incorrect positioning of a waterstop (or no waterstop where one is necessary)
  - incorrect positioning of a joint filler (or no filler where one is necessary)
  - failure of joggle joints by cracking
  - poor preparation of concrete between pours, i.e. scabbling (which may shatter aggregate and cause microcracking of the concrete) instead of washing, grit blasting or water jetting
- joints opening beyond the normal operating range of the sealants (indicating (irreversible) movement of the structure; checks should be made for subsidence, thermal effects and loading (as causes))
- failure of jointing materials including
  - deterioration of joint fillers and sealants ~ through age embrittlement, shrinkage (particularly in mastics), biodeterioration and biodegradation or chemical attack
  - splitting of waterstops ~ due to embrittlement or over extension
  - biological or chemical deterioration of the substrate material
- biodeterioration of joints and materials including
  - worms penetrating puddled clay and softening it
  - root entry into expansion joints or sliding joint material
- detritus jamming joints.

*Effects*
- leakage inwards or outwards from the reservoir
- cracking of the construction materials.

### A2.4.6 Breakdown of waterproofing systems

*Relevant construction groups*

| Roof | Wall | Column | Floor |
|---|---|---|---|
| All | All | None | All |

*Causes*
- failure of materials
- aggressive environments
- mechanical damage.

> For details see Appendix I, Section AP I.5.

*Effects*
- leakage inwards or outwards from the reservoir.

## A2.4.7 Problems with access (including access ladders), ventilation, openings and ancillary items

*Relevant construction groups*

| Roof | Wall | Column | Floor |
|------|------|--------|-------|
| All  | All  | None   | All   |

*Causes*
- entry of contaminants through ventilators
- corrosion of metalwork
- vandalism of the ventilators
- contamination of the water by insertion of items through the ventilators
- contamination of water by deterioration of materials (e.g. timber depth gauges)
- cracking of the surrounding concrete
- inadequate ventilation
- settlement/ground movement around connecting pipes
- dissociation of components e.g. upstands, metal roof fittings leaving a path for infiltration.

> For details of corrosion of metalwork see Appendix I, Section AP I.1.

*Effects*
- damage to ventilators
- contamination of water
- leakage inwards or outwards
- failure of roof/wall joint
- corrosion and failure of pipework (particularly at outlets)
- corrosion and failure of access ladders.

# A3 Methods of repair and material categories

> Considerations in the decision making process which should be read in conjunction with this Section are: Sections B1, B2, B3, B4 and B5.
>
> Information aiding the decision making process relevant to this Section can be found in all Sections of Part C.

The objective of this section is to identify the most appropriate methods and materials for waterproofing and repair.

In general, the optimum procedure in all cases of roof repair and waterproofing should be to carry out the work on the outside of the structure. However, in some cases this may not be possible. It is unlikely that repair from the outside will be feasible for walls or floors. Advice is therefore given on methods and materials for both internal and external repair.

Having placed the defect requiring repair into a **defect category**, as defined in Section A2, the following steps should be followed:

1. Select the most appropriate **method of repair**.

> In order that the most appropriate remedy is selected, the aims of the repair must be clearly defined: e.g. to restore structural or waterproofing integrity, to restore the original profile or to arrest deterioration. The most appropriate method of repair may be determined as much by practical and operational considerations as by technical criteria.
>
> A factor that may influence the choice of method of repair for roofs is the type of insulation that is to be used (see Section B4.1).
>
> For each **defect category**, a method statement is given in this Section which outlines the numerous methods of repair available and gives criteria for selection.

2. Identify the most suitable **material category**.

> For each **method of repair**, relevant **material categories** are identified. Information relating to the physical properties, use and approvals (to meet the water quality requirements, etc.) of each, is given in Sections A4 and C1.3.1.

The principles of each **method of repair** for the **defect categories** identified, and listed below, are given by means of a flow chart, with related advice given in the shaded boxes.

- A3.1 Corrosion of structural metalwork
- A3.2 Concrete and brickwork deterioration
- A3.3 Cracking (other than formed joints)
- A3.4 Corrosion of reinforcement
- A3.5 Joint failure
- A3.6 Breakdown of waterproofing systems
- A3.7 Problems with access (including access ladders), ventilation, openings and ancillary items

## A3.1 CORROSION OF STRUCTURAL METALWORK

```
                                              ┌─────────────────────────────────────────────┐
                                              │ Establish cause and extent of corrosion     │
                                              └─────────────────────────────────────────────┘
                                                                    ↓
┌─────────────────────────────────────────┐   ┌─────────────────────────────────────────────┐
│ This may require the replacement of     │   │ Remedy the cause of the corrosion           │
│ some elements or the design of a        │   │ (where possible)                            │
│ suitable insulation system where        │   │                                             │
│ bimetallic corrosion is occurring.      │   │                                             │
└─────────────────────────────────────────┘   └─────────────────────────────────────────────┘
                                                                    ↓
                                              ┌─────────────────────────────────────────────┐
                                              │ Assess structural adequacy of corroded      │
                                              │ member                                      │
                                              └─────────────────────────────────────────────┘
                                                                    ↓
                                              ┌─────────────────────────────────────────────┐
                                              │ Erect necessary temporary support and       │
                                              │ replace or strengthen structurally          │
                                              │ inadequate members                          │
                                              └─────────────────────────────────────────────┘
                                                                    ↓
                                              ┌─────────────────────────────────────────────┐
                                              │ Expose all metalwork to be painted          │
                                              └─────────────────────────────────────────────┘
                                                                    ↓
                                              ┌─────────────────────────────────────────────┐
                                              │ Remove oil, grease and chemical salts by    │
                                              │ high pressure water jetting                 │
                                              └─────────────────────────────────────────────┘
                                                                    ↓
```

| | |
|---|---|
| Preparation should be as appropriate for the metalwork paint system. The abrasive medium should be checked for oil by placing it in a vial of water, shaking and testing for the formation of a surface sheen.<br><br>All dust, etc. should be removed after treatment by vacuuming or some other equally effective method. Cleanliness can be assessed by the methods specified in BS7027: Group B (ISO 8502) adequate removal of dust can be checked using an adhesive tape method[4].<br><br>The surface profile of the blast-cleaned metal should have a peak-to-trough amplitude appropriate for the paint system to be applied. | Blast-clean to the required surface finish in accordance with BS7079: Part A1 :1989 |
| The surface of metalwork to be painted should be kept dry (to avoid the growth of further corrosion products) and should be primed (with an approved primer) within 4 hours of cleaning. Where it is not possible to keep the surface dry, then an approved moisture tolerant paint system should be used and the manufacturer consulted for advice on application.<br><br>The approved metalwork paint system should be applied by brush, roller or spray as recommended by the manufacturer. Each coat of paint should be applied to achieve the minimum dry film thickness requirement, with a uniform continuous layer. Alternatively, where appropriate, metal surfaces, e.g. roof joists, may be protected by a layer of sprayed concrete. | Apply METALWORK PAINT SYSTEM in accordance with manufacturers recommendations. |
| A period of forced ventilation may be necessary to remove all traces of the paint solvent from the reservoir and allow adequate curing, to ensure the water quality requirements can be achieved. | Cure paint system in accordance with the manufacturers recommendations. |

## A3.2 CONCRETE AND BRICKWORK DETERIORATION

```
                                    ┌─────────────────────────────────────┐
                                    │ Establish cause and extent of       │
                                    │ deterioration                       │
                                    └─────────────────┬───────────────────┘
                                                      ↓
                                    ┌─────────────────────────────────────┐
                                    │ Remedy the cause of the deterioration│
                                    │ (where possible)                    │
                                    └─────────────────┬───────────────────┘
                                                      ↓
```

| As a result of the structural assessment, weight restrictions may be placed on a roof, making any external construction activity impossible. | Assess structural adequacy of affected elements (see Section A3.2.1) |

↓

| | Erect any necessary temporary support and replace or strengthen inadequate members |

↓

| **Brickwork**<br><br>Flaking or delaminating brickwork should be blast-cleaned to remove all damaged material and then pressure washed and left to dry. A dehumidifier should then be used to draw moisture from the bricks. Unacceptable areas of damage should then be made good using a rendering material (which may be similar to a concrete repair material) and the surface protected by the application of a suitable surface coating[5].<br><br>An alternative to repair using a render, where the deteriorated areas are small, cut out and replacement with new bricks should be considered. | Repair using an appropriate CONCRETE REPAIR SYSTEM if section needs reinstating. |

↓

| It should be noted that surface coatings, like all repair systems have only a finite life and will require reapplication at intervals. | Apply a SURFACE PROTECTIVE COATING or WATERPROOFING MEMBRANE, as appropriate, if:<br><br>• only the surface of the concrete or brickwork is affected and further deterioration can be prevented, for example from:<br>  - chemical attack<br>  - physical attack<br><br>• protection to repair material is required, for example from:<br>  - chemical attack. |

### A3.2.1 Assessment of mass concrete and arched roofs

Reservoir roofs have different structural configurations, materials and level of deterioration: hence they will not show similar structural response during any loading/unloading operation. Prior to undertaking any repair work to an arch roof, an assessment of its behaviour should be conducted (see Section C2).

The removal of soil can lead to horizontal and vertical movements, the magnitude of which depends on the amount of load relieved, the sequence of soil removal and the type of roof under consideration. The removal of a large depth of soil from the roof can be sufficient to cause cracking at the crown of the arch (due to vertical movement) and at the roof/wall joints (due to horizontal movement). The sequence of soil removal is very important, especially in multi-span arch roofs. It may be least damaging to remove the soil uniformly over the whole roof rather than over successive single arch spans. The horizontal thrust resulting from the load retained on adjacent spans can substantially increase the horizontal and vertical (upwards) movements of an arch span relieved of the soil loading. Storage of soil removed from one span on an adjacent span can further increase the damage.

Removal of insulation will result in heating of the roof. The amount of expansion and possible cracking will depend on the roof structure, the sequence of removal of insulation and the duration of exposure (see Sections B3 and B4).

Construction plant and stored materials may subject the roof to sufficiently large local stresses to cause unacceptable cracking or even collapse. The type of plant and the working procedure to be adopted, storage of materials and the sequence of operations will depend on the structure under consideration. It is recommended that a careful assessment of the structural response to the possible scenarios of soil removal, insulation removal and working operations (plant and materials) be undertaken prior to any repair work. Such an assessment would require knowledge of the structural details (dimensions and materials) of the roof. If records of these are not readily available, a preliminary survey would be necessary to obtain data for the assessment.

There are a number of levels of analysis that can be performed for the assessment of arch structures. These would include; the elementary line of thrust method, the plastic method and advanced computer methods. These methods vary in the degree of complexity, accuracy and the details of structural response obtainable. The level of assessment warranted by a particular structure should be determined from its geometrical details.

For stocky arches (span/depth ratio less than 15), tensile stresses are unlikely to develop, and linear elastic behaviour can be expected. Linear analyses using elementary methods[6] can be performed by hand or a simple frame analysis program may be used. Arches with a span/depth ratio exceeding 15 will be flexible and have a non-linear response. Tensile stresses can develop, resulting in cracking and thinning of the arch structure. These structures need to be analysed using computer programs or other advanced methods. Software suitable for this type of analysis would include thinning elastic methods[7] and mechanism methods[8, 9].

The thinning elastic method can perform analyses of the arch structure to provide details of the non-linear responses at serviceability and collapse states. Stresses, displacements, load paths, development of cracking and 'thinning' of the arch are readily computed for different load combinations. The mechanism method of analysis can perform an analysis of the collapse state only and hence is less versatile than the thinning elastic method. Computer packages using these methods of analyses are readily available.

> However, experience in the use of such programs is necessary in order to carefully model the structure, select the input parameters and interpret the resulting structural response. Such a task would be best conducted by specialists.

## A3.3 CRACKING (OTHER THAN FORMED JOINTS)

| | |
|---|---|
| In general, cracks that are parallel to the main reinforcement are likely to be related to reinforcement corrosion. Cracks at right angles or diagonal to the main reinforcement are more likely to result from structural loading or movement (see Appendix I Section AP I.3). It is advisable that all cracks in reinforced concrete be sealed to prevent corrosion of the reinforcement and contamination of the water (see Section C1). When a liquid membrane is to be applied the sealing of cracks over a certain width to assist bridging of the membrane should also be considered. | Establish cause and extent of cracking and remedy cause. |
| As a result of the structural assessment, weight restrictions may be placed on a roof, making any external construction activity impossible. Care must be taken when carrying out surface preparation, e.g. cutting out cracks for repair, to prevent stresses being induced that are sufficient to further damage the roof. | Assess cracked elements for structural adequacy and strengthen if necessary. |
| When crack sealing is being relied upon to prevent leakage into the reservoir, it is recommended that sealed cracks be water tested by ponding prior to the installation of the waterproofing membrane as specified in the Civil Engineering Specification for the Water Industry (CESWI Clause 7.14).<br><br>Use a de-bonding strip over moving cracks. | **External repair**<br><br>Lay WATERPROOFING MEMBRANE on the outside:<br><br>• As a priority in a repair contract brought about by leakage into the reservoir.<br>• Either with or without crack sealing. |
| If water egress has resulted in backfill or ground erosion then fill resultant voids by LARGE VOLUME GROUTING. This method may also be used to seal the cracking. | **Internal repair**<br><br>Seal cracks from the inside of the reservoir if:<br><br>• Access to top of roof or outside of walls is not available.<br>• Leakage is out of the reservoir.<br>• Leaks are through the floor. |
| | Apply a WATERPROOFING TREATMENT to the inside if:<br><br>• There are a large number of fine cracks, e.g. through joints in brickwork. |
| Guidance on how to test for movement is given in Section C2. If this is difficult then all cracks should be treated as moving. | Monitor cracks for movement. |
| | Cracks should be filled by CRACK INJECTION where:<br><br>• Cracks are detrimental to structural integrity and continuity is required.<br>• Cracks are inaccessible.<br>• Cracks are not moving. |
| The term 'crack sealant' is defined as any material that can be used to seal a crack and can be one of the following three types:<br>• Non-flexible sealants.<br>• Flexible sealants.<br>• Preformed sealants. | Cracks up to 1 mm wide which are not moving should be sealed using a NON-FLEXIBLE SEALANT, as shown in Figure A3.1. |
| | Cracks greater than 1 mm wide which are not moving should be sealed by CRACK FILLING. |
| If a flexible crack sealant is used then it is desirable to make the seal considerably wider than the crack in order to limit the strain to that which the sealant can accept. | All other cracks that are moving should be sealed with either a FLEXIBLE SEALANT, see Figure A3.2a, or a PREFORMED SEALANT, see Figure A3.2b and A3.2c. |

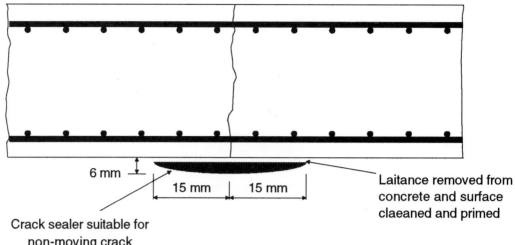

**Figure A3.1** *Sealing of non-moving cracks up to 1 mm wide*

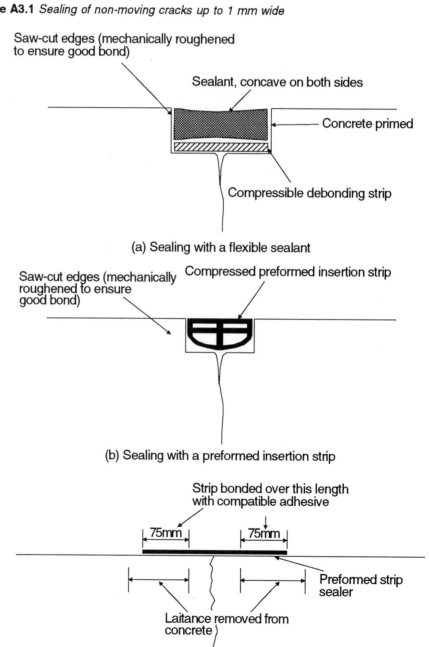

(a) Sealing with a flexible sealant

(b) Sealing with a preformed insertion strip

(c) Sealing with a preformed strip sealer

**Figure A3.2** *Methods of sealing moving cracks*

## A3.4 CORROSION OF REINFORCEMENT

Deterioration of concrete, where protection to reinforcement is not lost and where it is considered that there is an adequate key for any repair material, should be treated by the method described for mass concrete in Section A3.2. This Section therefore deals specifically with deterioration leading to corrosion of the reinforcement.

---

Establish cause and extent of reinforcement corrosion and remedy cause (if possible).

↓

| | |
|---|---|
| As a result of the structural assessment, weight restrictions may be placed on a roof, limiting the feasible repair methods. | Estimate loss of reinforcement cross-section and assess affected members for structural adequacy (refer to Sections AP I.4 and C2). |

↓

Prior to the removal of any concrete, erect necessary temporary support

↓

| | |
|---|---|
| The investigation and repair of prestressed and post-tensioned reservoirs is a specialist task and should only be undertaken by experienced personnel.<br><br>Specialist advice should be sought if steel plate bonding is being considered. | If member is structurally adequate then:<br><br>• Thoroughly clean corroded steel.<br>• Reinstate section with CONCRETE REPAIR SYSTEM.<br><br>If member is structurally inadequate then:<br><br>• Couple-in additional/replacement reinforcement.<br>• Replace member.<br>• Bond on steel plates. |

↓

| | |
|---|---|
| If the concrete is carbonated to within 5 mm of the reinforcement it is recommended that repairs should be carried out.<br><br>When corrosion is due to chloride attack (e.g. from use of filter bed gravels or unwashed sea dredged aggregate), then all concrete with a chloride content of more than 0.3% by weight of cement (Department of Transport Departmental Advice Note BA 23/86) must be removed.<br><br>It is also essential that thorough cleaning of any reinforcement to be retained be carried out. Wire brushing is usually inadequate where chloride induced corrosion has occurred[10] and grit blasting or water jetting (for fine pits) to Sa2½ in accordance with BS7079: Part A1: 1989 are recommended.<br><br>The use of cathodic protection, desalination or re-alkalisation techniques are not considered appropriate for most service reservoirs; specialist advice should be sought before their use. | If corrosion is due to chemical attack then:<br><br>• Remove affected concrete to a minimum distance of 20 mm behind the reinforcement; refer to Section A4.2.1.<br>• Remove or thoroughly clean corroded steel and couple in new bars.<br>• Reinstate section with CONCRETE REPAIR SYSTEM. |

↓

| | |
|---|---|
| An example of where a coating may be sufficient is the corrosion (without spalling) of expanded mesh reinforcement, used in roofs built at the turn of the century. Rust stains will usually be visible on the soffit, and although trial areas of concrete should always be removed to assess the reinforcement condition, it may be that adequate protection to the reinforcement can be provided by increasing the depth of the cover (by sprayed concrete or other suitable overlay) or by applying a SURFACE COATING on the inside. | If corrosion is due to inadequate concrete cover or porous concrete then:<br><br>• Remove concrete until a sound substrate is reached to a minimum depth of 20 mm behind the reinforcement.<br>• Reinstate section with CONCRETE REPAIR SYSTEM<br>• Alternatively, a SURFACE COATING may be sufficient. |

## A3.5 JOINT FAILURE

Establish cause and extent of joint failure and remedy cause if possible.

---

BS6213 *'A Guide to the Selection of Construction Sealants'* gives guidance on the joint width to be taken for design purposes and on the selection of the most appropriate joint sealant. Joint design requires the calculation of the expected thermal movement, which is carried out as follows. Initially, identify the position and effectiveness of restraints to movement such as cross-walls and columns. Then assess the length of concrete for which the joint is to provide movement relief, $L$ (mm). Secondly, determine the temperature range to which the concrete will be exposed, $T$ (°C). The expected total movement, $M$ (mm), can then be calculated from:

$$M = A \times L \times T$$

where $A$ = the coefficient of thermal expansion of concrete taken as $10 \times 10^{-6}$ /°C.

Alternatively, a calculation can be made of the expected expansion or contraction for any given temperature range on the assumption that a seal will be installed while the concrete is in equilibrium at a known current temperature.

Concrete expansion $= A \times L \times (T_{max} - T_c)$
Concrete contraction $= A \times L \times (T_c - T_{min})$

where   $T_c$       = Current temperature
        $T_{max}$   = Maximum expected temperature
        $T_{min}$   = Minimum expected temperature

Calculate expected joint movement to enable suitable repair method and materials to be selected.

---

The repair of wall and floor joints may make this impractical or uneconomic. The application of a WATERPROOFING TREATMENT on the inside, localised waterproofing using a BONDED or LIQUID MEMBRANE, or specific JOINT SEALING may then be considered.

When joint sealing is being relied upon to prevent leakage into the reservoir, then it is recommended that sealed joints be water tested, by ponding, prior to the installation of any waterproofing membrane, as specified in the Civil Engineering Specification for the Water Industry.

Repair of roof/wall joint

Failure can be exacerbated by significant thermal, or other, differential movement between the roof and wall. The removal of the roof insulation for repair will often increase or create this movement. The joint could be repaired by any of the methods described below; alternatively, if a bonded membrane is used, care should be taken to provide for movement of the joint.

Lay WATERPROOFING MEMBRANE on the outside:
- As a priority in a repair contract brought about by leakage into the reservoir through the roof.
- Where access is available to the outside of walls and contamination of the reservoir is a problem.
- Where leakage is through the roof/wall joint.
- Either with or without joint sealing.

---

The principal problem with sealing joints from the inside of a reservoir roof is that the hydrostatic pressure from ground or surface water entering the reservoir is acting to force the sealant out of the joint.

Repair from inside the reservoir if:
- Access to the outside of roof or walls is not available.
- Leaks are through the floor.
- Leakage is out of the reservoir.

|  | If water egress has resulted in backfill erosion then the resultant voids should be filled by LARGE VOLUME GROUTING. This method may also be used to fill failed joints. |
|---|---|
| Care should be taken not to grout the joint itself. | INJECTION TECHNIQUES can be used to repair joints when:<br>• Concrete adjacent to the joint has failed, e.g. around a waterstop |
| New WATERSTOPS should always be welded *in situ* to existing WATERSTOPS to ensure continuity. | A CONCRETE REPAIR SYSTEM can be used when:<br>• The damage to adjacent concrete is too great to seal.<br>• It is necessary to replace a failed WATERSTOP. |

Remove failed JOINT SEALANT.

Joints should then be sealed with either a JOINT SEALANT or a PREFORMED STRIP SEALANT.

Where joint failure is due to filler failure, for example stones becoming trapped in the joint or bacteriological problems, then the JOINT FILLER must be replaced with one designed for the purpose. The joint should then be sealed with a JOINT SEALANT.

## A3.6 BREAKDOWN OF WATERPROOFING SYSTEMS

```
┌─────────────────────────────────────────────────┐
│ Establish cause and extent of breakdown and     │
│ remedy cause (if possible).                     │
└─────────────────────────────────────────────────┘
                        ↓
┌─────────────────────────────────────────────────┐
│ Establish purpose and necessity of replacing    │
│ membranes and consider suitability of other     │
│ methods of repair.                              │
└─────────────────────────────────────────────────┘
                        ↓
┌──────────────────────────────┐  ┌───────────────────────────────────┐
│ See Section B2. Where a roof │  │ If poor drainage has contributed  │
│ has insufficient fall a      │  │ to the failure consider upgrading │
│ SCREED should be laid.       │  │ the drainage system as part of    │
│                              │  │ the repair.                       │
└──────────────────────────────┘  └───────────────────────────────────┘
                                                  ↓
                                  ┌───────────────────────────────────┐
                                  │ Repair existing WATERPROOFING     │
                                  │ MEMBRANE if:                      │
                                  │                                   │
                                  │ • Existing membrane is largely    │
                                  │   intact and has adequate         │
                                  │   residual life.                  │
                                  │ • Compatible products exist.      │
                                  │ • Economy is required.            │
                                  └───────────────────────────────────┘
                                                  ↓
┌──────────────────────────────────────┐ ┌───────────────────────────────────┐
│ The installation of a new membrane   │ │ If the existing membrane cannot be│
│ will require consideration of the    │ │ repaired then a new membrane must │
│ role of the existing membrane.       │ │ be installed.                     │
│ There are two possible options:      │ │                                   │
│                                      │ │ Lay new WATERPROOFING MEMBRANE    │
│ 1. Retain in place, i.e. lay new     │ │ over existing if:                 │
│    membrane over the existing.       │ │                                   │
│                                      │ │ • Existing membrane and primer    │
│ 2. Remove the existing membrane.     │ │   are not easily removed.         │
│                                      │ │ • Surface of existing membrane is │
│ Wherever possible, the existing      │ │   suitable to receive a new       │
│ membrane should be retained in       │ │   membrane.                       │
│ place; however, the removal of an    │ │ • New membrane is chemically      │
│ existing membrane will depend on     │ │   compatible with existing, in    │
│ the particular circumstances and may │ │   the case of liquid applied or   │
│ be a very complex decision.          │ │   bonded membranes.               │
│                                      │ │                                   │
│ Removal of the existing membrane     │ │ Remove existing membrane and      │
│ may only be likely if a new BONDED   │ │ replace with a new WATERPROOFING  │
│ MEMBRANE is required which may also  │ │ MEMBRANE if:                      │
│ require a SCREED to be placed.       │ │                                   │
│                                      │ │ • existing membrane is not intact │
│                                      │ │   and/or                          │
│                                      │ │ • existing membrane and primer    │
│                                      │ │   can easily be removed.          │
└──────────────────────────────────────┘ └───────────────────────────────────┘
                                                  ↓
                                  ┌───────────────────────────────────┐
                                  │ Repair using alternative          │
                                  │ techniques such as CRACK          │
                                  │ INJECTION or JOINT SEALING if     │
                                  │ considered more appropriate       │
                                  └───────────────────────────────────┘
```

> It is recommended that, where practicable, membranes be water tested by ponding, as specified in the Civil Engineering Specification for the Water Industry clause 7.14[11], prior to being covered. On reservoirs that slope, it is more practical to pond test small areas at any one time (see Figure A3.3). It is preferable for ponding tests to be undertaken over a period of several days to check the membrane for watertightness, allowing time for leakage water to find defects in the roof. Where this is impracticable then a hosing test is acceptable.

**Figure A3.3** *Pond testing a small area on a reservoir roof*

## A3.7 PROBLEMS WITH ACCESS (INCLUDING ACCESS LADDERS) VENTILATION, OPENINGS AND ANCILLARY ITEMS

Problems with access, openings and ventilation fall into several different categories:

1. The access or ventilator design not preventing the inward passage of sources of contamination.

> This can sometimes be solved by the use of approved seals and the insertion of a fine mesh made of a non-corroding material into the air gap, or by redesigning and reconstructing the vent or access opening.
>
> Gas struts or torsion springs can provide lift assistance, enabling single-man operation, and safety grids can be incorporated to prevent accidental entry or lid closure.
>
> The number of ventilators provided on the roofs of older reservoirs is often surplus to requirements, and remedial works may therefore include the removal of a number. Further details of ventilation requirements are given in Section A3.7.1.
>
> Where access covers are to be replaced their design should be checked to ensure no loss of ventilation capacity of the reservoir will occur.

2. Cracking in the surrounding concrete.

> This should be dealt with either by CONCRETE REPAIR SYSTEMS or by CRACK SEALING, after establishing and remedying the cause, if possible. If it is not possible to treat the cause then materials should be selected that are tolerant to the conditions.

3. Damage from vandalism.

> Vents and access covers could, depending on the extent and frequency of vandalism, be rebuilt with vandal resistant materials and improved security considered. Concealed mounting bolts, hasps and hinges and internal retention brackets can be incorporated to improve security. Micro switches can also be fitted to detect intruders. Examples of vandal resistant designs are shown in Figures A3.4 to A3.6.
>
> GRP is not considered to be a vandal resistant material, as it may be severely damaged by fire.
>
> Where access covers are to be replaced their design should be checked to ensure no loss of ventilation capacity of the reservoir will occur.

4. Corrosion of metalwork.

> Corrosion of metalwork in vents and access covers can usually be treated with a METALWORK PAINT SYSTEM (see Section A4.2.6).
>
> Where vent corrosion is severe, it may be more economical to replace the vent. The replacement vent selected should be resistant to corrosion: e.g. an appropriate grade of stainless steel or suitable plastic.
>
> Bimetallic corrosion in the welds of rungs on access ladders is a safety hazard and usually requires replacement of the ladder with one made of non-corroding material; see also Appendix I, Section AP I.1.

**Figure A3.4** *Vandal resistant access design*

**Figure A3.5** *Vandal resistant combined access and vent*

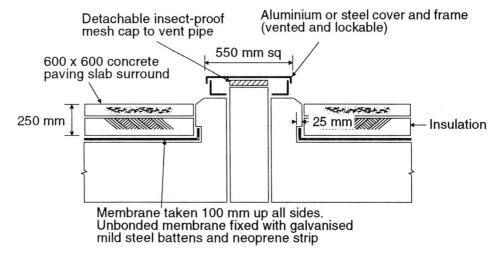

Detail A: Vent with brick / concrete surround

Detail B: Vent with vented and lockable roof cover

**Figure A3.6** *Typical details for vandal resistant vents*

CIRIA Report 138                                                                 A27

5. Contamination of water by other materials.

> Where materials such as lead (formerly used for repairing joints) or timber (formerly used for access ladders or depth gauges) are present inside the reservoir they should be removed and replaced with items manufactured from appropriate materials

6. Failure of pipework.

> Remedial works to pipework that has failed due to corrosion or ground movement are generally not possible and the affected sections will need to be replaced.

7. Dissociation of roof components.

> Dissociation of roof components, such as metal roof fittings and upstands, leaving a path for infiltration and contamination.

### A3.7.1 Ventilation requirements for reservoirs

The need for adequate ventilation for reservoirs to avoid the build-up of harmful or explosive gases is briefly mentioned in BS8007: 1987 (Cl.2.9.3). Ventilators in reservoir roofs also have two other functions: to allow the free circulation of air above the stored water, preventing its stagnation, which could result in the build-up of unacceptably high chlorine levels; and to prevent loading of the roof from suction during drawdown (downward forces) or pressure increase (upward forces) during filling. Where a roof is continuous over a dividing wall the combined effects of suction on one side (during emptying) and upward pressure on the other (from filling) should be considered. Problems are likely to be exacerbated where the rate of emptying or filling is fast.

Traditionally the number and size of vents has been determined empirically or by 'eye', and may have resulted in more ventilators being provided than are necessary. In older reservoirs large numbers of ventilators may have been provided for aesthetic reasons.

As each ventilator is a potential source of ingress of contaminants, it is desirable to minimise the number, while still retaining a sufficient number for adequate ventilation and to prevent 'implosion' on drawdown. It should not be assumed that leakage (of air) through the wall/roof joint and or around access openings will relieve the suction or pressure effects. These should not be considered as part of the ventilation system when the number of ducts or area required is calculated. Likewise, the area of any overflow pipes should only be allowed for in the calculation if the designer is certain that they will not become blocked or covered (e.g. by flaps). The aim should be to provide at least one ventilator in each bay of a roof: i.e. the area between beams, which may be isolated if the water level rises above the beam soffit.

In practice, sufficient vents should be present to ensure that the pressure differential developed during drawdown or filling does not exceed the value assumed in the design. This factor can then be checked during the structural assessment of the reservoir.

> In cases where the value is not known, a maximum pressure differential must be assumed by the Engineer for the particular structure.

The number of vents required can be calculated from:

$$N = 1.11 \frac{k_o}{k_i} \sqrt{\frac{2gH}{\Delta P}}$$

for the general case or, where a sudden water loss (at the maximum rate) may occur, e.g. a burst pipe, etc., from:

$$N = 4.91 \frac{k_o}{k_i} \sqrt{\frac{H}{\Delta P}}$$

Where:    $k_o = C_{do}A_o$
          $k_i = C_{di}A_{std}$

and the following terms apply:

$A_o$ = Outlet area (m²).
$A_{std}$ ≡ The **effective** area of a standard vent.
$\Delta P$ = Differential pressure drop across the vent (N/m²).
$C_{do}$ = Coefficient of discharge of the outlet of the reservoir (see Appendix II, Note 3).
$C_{di}$ = Coefficient of discharge of the inlets of the reservoir (see Appendix II, Note 4).
$H$ = Maximum water depth (height of the reservoir) (m).

> Note that:
>
> - $\Delta P$ must be limited, by providing sufficient vents to prevent structural damage or, more importantly, to limit air velocity.
>
> - No factor of safety is assumed against vent failure or reduced operating capacity (a 20% reserve may be appropriate, or it could be considered during the design where $\Delta P$ is a factored pressure differential).
>
> The full derivation of the formula and an example calculation are given in Appendix II.

# A4 Repair materials

> Considerations in the decision making process that should be read in conjunction with this Section are: Sections B1, B5 and B6.
>
> Information aiding the decision making process relevant to this Section can be found in: Sections C1, C3, C4 and C5.
>
> There are a large number of repair and waterproofing materials on the market that may be suitable for reservoir work. Over the duration of the research culminating in this report most British suppliers were contacted together with a number of suppliers from overseas. However, there may be material types currently on the market or new materials that may be suitable for a particular application that have not been included. It is accepted therefore that other repair methods should not necessarily be excluded.

For each material category suitable materials are described by their generic group. The properties of different materials within a group can vary widely. **Example performance requirements** have been detailed in Section B6.

## A4.1 MATERIALS APPROVAL SCHEMES

It is a legal requirement that Water Undertakers comply with the requirements of Regulation 25 of the Water Supply (Water Quality) Regulations, 1989 (incorporated into the Water Industry Act, 1991[12]) when introducing any substance or product into water that is to be supplied for drinking, washing or cooking. Any contravention of Regulation 25 is a criminal offence under Regulation 28, which may lead to prosecution of the Water Undertaker.

There are other independent certification/approval schemes in use for repair and waterproofing products to assist in the selection of appropriate materials; however, certification under these schemes is not a legal requirement, unless specified in a contract document. These include, the Water Byelaws Scheme (WBS), British Board of Agrément (BBA) certification and Water Industry Certification Scheme (WICS, now operated by BSI). WICS is only one of several certifying bodies involved in third-party certification of water supply products. The way that these schemes should be taken into account, when selecting materials for use in contact with potable water is summarised below.

### A4.1.1 Regulatory Approvals System

The Water Industry Act, 1991 (which consolidated and incorporated the Water Act, 1989[12]) gave the Secretary of State for the Environment and the Secretary of State for Wales power to regulate the control of substances, products and processes in the treatment and provision of water supplies in England and Wales. The relevant provisions (Regulations 25 (products) and 26 (processes)) of the Water Supply (Water Quality) Regulations 1989 (amended 1989 and 1991)[13] came into force in September 1989. Similar regulations apply in Scotland.

Regulation 25.-(1), requires that a Water Undertaker shall not, otherwise than for the purposes of testing or research, apply any substance or product to, or introduce any substance or product into water which is to be supplied for drinking, washing and cooking, unless it:

(a) has been approved by the Secretary of State (Regulation 25 (1) (a)); or
(b) is considered by the Water Undertaker to be unlikely to affect adversely the quality of the water (Regulation 25 (1) (b)); or
(c) has been used by a Water Undertaker during the 12 months prior to 6 July 1989 (Regulation 25 (1) (c)); or
(d) is listed in the 15th statement or any supplement issued by the Committee on Chemicals and Materials of Construction for use in Public Water Supply and Swimming Pools (Regulation 25 (1) (d)).

Regulation 25 (8) requires that the Secretaries of State issue, at least once each year, a list of all substances and products for which approval has been granted under regulation 25 (1) (a). Since 1 September the DoE-CCM has advised the Secretaries of the Environment, Wales and Scotland on applications for approval made by manufacturers and on matters relating to the revocation, modification of approvals and prohibition of substances or products.

The current approval system is being reviewed by the Drinking Water Inspectorate (dwi) to take account of EC Directives and moves towards European standardisation.

Materials listed in the statements and supplements of the DoE-CCM are approved by virtue of Regulation 25 (1) (d). For approval under this scheme materials will have been assessed for the potential for long- and short-term risk to the health of the consumer arising from the use of the product in contact with potable water. Approval is subject to certain conditions, e.g. specified method of application or treatment after application.

> Approval of a material under Regulation 25 (1) (a) and (d) will be for a specific composition and will require that the material be retested, for approval and listing, every time a change in the formulation is introduced.
>
> Care should be taken to ensure that a material selected for use under Regulation 25 (1) (a) and (d) has the same formulation as that approved by the Secretary of State.

### A4.1.2 Water Byelaws Scheme (WBS)

The WBS tests water fittings (including materials) that are to be used after the time of supply (i.e. when the water has passed from the Water Undertaker's pipe into the consumer's pipe) to establish whether they satisfy the requirements of the water byelaws, which require that installers and consumers do not permit contamination, cause waste or misuse of public water supplies. For most non-metallic products the methods and criteria specified in BS6920[14] (which apply solely to the effect of materials on the quality of potable water with which it may come into contact) are used. Five water quality tests are specified by BS6920: taste of water, appearance of water, growth of aquatic microorganisms, the extraction of substances that may be of concern to public health and the extraction of metals. All tests are undertaken on test panels of the material which have been applied and fully cured in accordance with the manufacturers recommendations. The components of the material may not, individually, satisfy the WRc requirements. Materials having passed these tests are listed in the Water Fittings and Materials Directory (WRc listed) and are referred to as Water Byelaws Scheme (WBS) approved products.

It has been assumed for this Manual that WBS approved products (as having satisfied the requirements of BS6920) are the **minimum** standard acceptable for use for the repair and waterproofing of service reservoirs.

> In addition, the requirements of the Control of Substances Hazardous to Health (COSHH) Regulations 1988, Consumer Protection Act 1987 and Health and Safety at Work Act 1974 should be specified in all contracts.

### A4.1.3 British Board of Agrément (BBA)

Agrément is the process of independent performance-based assessment taking regard of information that already exists or has to be obtained. It is not a standard, as standards do not recognise distinctions between materials. The BBA examines laboratory test data, site experience, production surveillance and suitability for the intended purpose. The possibility of contact with potable water is not taken into account.

### A4.1.4 Water Industry Certification Scheme (WICS)

This scheme operates with the support of the UK water industry, and it offers both certification of quality management systems and product conformity.

## A4.2 MATERIAL CATEGORIES

The following sections define the material categories that have been identified in Section A3 as follows:

- concrete repair systems

- screeds

- waterproofing membranes on the outside
  - bonded
  - unbonded
  - unbonded hydrophilic membranes
  - liquid

- internal waterproofing treatments

- surface protective coatings

- metalwork paint systems

- crack injection

- crack fillers and sealants
  - non-flexible
  - flexible
  - preformed
  - crack filling

- joint fillers and sealants
  - joint fillers
  - joint sealants
  - preformed strips

- waterstops

- large volume grouting
  - cementitious
  - resin.

The various structural elements in reservoirs are subjected to different exposure conditions and therefore the same waterproofing and repair materials and systems may not be appropriate for all elements. For example, the roof is subjected to environmental effects and the complications resulting from wetting, drying and ponding. Walls are governed by changes in level of the water table, water rising by capillary action from the ground, etc., on the outside and the continual changes of water level on the inside. Floors are permanently under water.

Many old reservoirs were constructed above natural springs and consequently have a high surrounding water table, which has often led to problems of water ingress. However, this is not generally a problem for new reservoirs.

### A4.2.1 Concrete repair systems

Selection of the most suitable concrete repair system will depend on the size, depth and location of the area to be repaired.

> Compatibility between the repair system and the substrate concrete, particularly for shrinkage, is of great importance. This may be overcome by the use of aggregate (with dimensions as large as the individual repair will permit) or additives in repair mortars. Mortars, used in wall repairs, should only be placed up to a maximum depth of 50 mm, in layers not exceeding 25 mm thick. For repairs with a depth greater than 50 mm, concretes (with minimum coarse aggregate size of 10 mm) should be used.

The possible components to a concrete repair system are reinforcement primer, bonding coat, the repair material and any subsequent render or coating plus any metal cover strips over joints if required.

> The components should be used only when they are an integral part of the repair system and when recommended by the manufacturer for use in a particular instance.

*(a) Reinforcement primers*

Reinforcement primers provide chemical, sacrificial or barrier protection to the reinforcement.

The most common primer generic groups are:

- cementitious

> Usually polymer modified and may also act as a bonding coat; see Section A4.2.1.(b). Brush application ensures the entire surface of the bar is coated with cementitious material, eliminating the possibility of sections of the bar not being passivated by air voids trapped at the reinforcement/repair material interface.

- zinc-rich primers

> These are designed to corrode sacrificially and the zinc content is crucial for their successful operation. They therefore require thorough mixing before and during application. The reinforcement requires a high standard of surface preparation: grit blasting to Sa2½ in accordance with BS7079: Part A1: 1989 is usually recommended.

- epoxy barrier coats.

> These are recommended for patch repairs where high levels of chloride are expected to remain in the vicinity of the reinforcement. They require a high standard of surface preparation (grit blasting to Sa2½ in accordance with BS7079: Part A1: 1989), prevent the formation of a passive layer around the reinforcement, and may reduce the bond between the reinforcement and the repair material unless aggregate is introduced onto the surface of the coating (blinded). Some formulations contain a proportion of cementitious material to provide an alkaline environment and stabilise the passive film. This is unlikely to occur as very little cement will be in direct contact with the reinforcement; the cement grains are encapsulated in the epoxy, which is highly impervious to water and oxygen, preventing hydration of the cement and growth of the passive film.
>
> This does not apply to the use of fusion bonded epoxy coated reinforcement in new construction, where the protective coating is applied under factory conditions with strict quality control procedures.

*(b) Bonding coats*
Where a bond coat is part of a proprietary repair system it should always be used and applied in accordance with the manufacturers instructions.

Where non-proprietary repair systems are to be used, bonding coats applied to the parent concrete are recommended when repairing all areas larger than 1 $m^2$ with an unmodified concrete mix or a concrete mix with a polymer additive. A bond coat would provide greater assurance of adhesion, if selected and applied correctly, for all repairs. The bonding coat should consist of cement, water, possibly fine sand and a modifying material in the form of a latex, which should be compatible with the polymer in the repair material and used only with the approval of the repair material manufacturer. Recommended bond coat modifying materials are:

- styrene butadiene rubber (SBR)
- acrylic.

> If bond coats are not used in accordance with the manufacturer's recommendations, are allowed to dry out before application of the repair material, are used on a very dense substrate, or are applied in too thick a layer, they may act as a bond breaker and reduce the bond to the substrate.

*(c) Repair material – horizontal surface*
Repair would usually involve the removal of concrete and its replacement by a suitable method. One of the following materials should be considered:

- a specially designed hand or trowel placed polymer modified cementitious material

> For soffit application, these are available containing lightweight aggregate. However, compaction behind reinforcement is difficult to obtain when working in an overhead situation, as is good bond to the substrate. A layer thickness of 25 mm should not be exceeded, to prevent sag.

- a polymer modified flowable concrete

> These can be either pumped or gravity poured behind a shutter. It is difficult to prevent voiding at the soffit interface with the use of flowable concretes.

- sprayed concrete

> This has the advantages of ease of construction and quicker completion times, although an experienced nozzleman is a necessity.
>
> Good compaction can be obtained with this method. However, one drawback is that for a flat soffit application these materials may need mesh reinforcement to be preplaced, although an exposed reinforcement bottom mat may suffice. The use of steel fibre reinforced shotcrete may overcome the need for a mesh and will also provide some structural capacity and resistance to cracking. For wall applications layers of up to 150 mm can normally be sprayed without the need for a reinforcement mesh, though they may need to be tied to the existing wall. The impermeability of sprayed concrete in thin layers can be improved by the addition of polymer emulsions.

- as an alternative to replacing damaged concrete a new concrete face could be cast or sprayed onto the existing concrete member. The minimum thickness should be 75 mm (though this may be reduced for non-structural applications with proprietary materials) and the overlay should be anchored to the substrate by a fixed mesh. The use of fibre reinforcement will provide some structural capacity and reduce cracking.

*(d) Repair material – vertical surface*
Any of the methods listed above for horizontal application would also be appropriate for application to a vertical surface. In addition, materials are available that are specifically designed for application to a vertical surface, i.e. those with thixotropic or flowing (for application within a shutter) properties.

> Lightweight mortars frequently specified for soffit application may not be suited for structural repairs owing to their reduced physical properties.

### A4.2.2 Screeds
Screeds may be required for two purposes:

- A screed of varying thickness can be placed over a roof to enhance the slope for drainage efficiency.

> Whenever a roof is being waterproofed then a slope of at least 1 in 100, and preferably 1 in 80, should be provided to prevent standing pools of water, which can greatly reduce the working life of some waterproofing membranes. It must be recognised that this may increase the loadings on a roof.

- A screed of uniform thickness can be placed over a roof or floor to provide a smooth surface suitable for the application of a bonded membrane or other surface treatment.

When used **for drainage**, a screed thickness of several hundred millimetres may be necessary. As reservoir roofs are designed for only a specific, and limited, imposed load then the screed usually has to be light in weight. The choice of material may depend on access limitations for plant on the roof, placing methods and the type of waterproofing membrane employed. Suitable materials would be:

- polystyrene, density approximately 20 kg/m$^3$

- lightweight aggregate, such as pelleted pulverised fuel ash, density approximately 1500 kg/m$^3$

- lightweight aggregate and cement, density approximately 1100 kg/m$^3$ (which requires a 13 mm cement–sand topping if a bonded or liquid membrane is to be used)

- no-fines lightweight aggregate and cement, density approximately 700 kg/m³ (mix 1 part cement to 8–10 parts lightweight aggregate, which requires a 13 mm cement–sand topping if a bonded or liquid membrane is to be used)

- foamed or aerated concrete, density usually between 500 and 1100 kg/m³.

When used to provide **a surface for a bonded membrane**, the screed will be relatively thin, and will require good adhesion to the substrate and a tensile strength sufficient to overcome cracking and de-bondment due to shrinkage and thermal stresses during construction. Suitable materials would be:

- cement and sand, density approximately 2000 kg/m³ (if only thin screed is required)

- no-fines aggregate and cement, density approximately 1800 kg/m³, which requires a 13 mm cement–sand topping if a bonded or liquid membrane is to be used

- foamed or aerated concrete, density usually between 500 and 1100 kg/m³

- fibre reinforced or polymer modified cementitious mortar or concrete, density approximately 2000 kg/m³

- epoxide resin, density approximately 2000 kg/m³.

> The minimum thickness for each of the above materials types will be different, and the recommendations of the manufacturer should be sought.
>
> Within thin screeds the possibility of reflection of cracks and joints through the screed exists.

Other applications for screeds may be: to contribute to the thermal insulation properties of the roof or to provide protection for a membrane. Cement and concrete screeds have a high thermal conductivity and will not significantly contribute to the thermal insulation.

### A4.2.3 Waterproofing membranes

Membranes can be applied either as liquids or as bonded and unbonded preformed sheets. Generic types include:

1) Bonded preformed membranes (see Section A4.2.3 (a))

- bitumenized fabric (hot or cold applied; solar reflective)
- elastomeric bituminous
- self-adhesive High Density Polyethylene (HDPE).

2) Unbonded preformed membranes (see Section A4.2.3 (b))

- butyl rubber
- Ethylene Propylene Diene Monomer (EPDM)
- chlorosulphonated polyethylene (Hypalon)
- Low Density Polyethylene (LDPE)
- reinforced Low Density Polyethylene (LDPE)
- High Density Polyethylene (HDPE)
- polyvinyl chloride (PVC).

3) Unbonded hydrophilic membranes (see Section A4.2.3 (c))

- sodium bentonite and geofabric
- sodium bentonite and paper.

4) Liquid membranes (see Section A4.2.3 (d))

- fibre-reinforced bituminous
- EPDM
- polysulphide
- polyurethane
- elastomeric bituminous
- silicone rubber
- glass reinforced polyester (GRP)
- elastomeric cementitious
- flexibilised epoxy systems.

Where there are a large number of cracks, the most practical method of sealing the surface is to cover it with a membrane. External treatments are favoured for roofs but are generally impracticable for remedial work to walls, as installation would necessitate excavation and possibly dewatering, leading to instability of the walls (see Section B4.2). Mastic asphalt was the material traditionally used for external tanking but is now no longer used.

Externally applied preformed sheet materials should be selected to withstand the expected movement, possibly cyclic, without damage. Bitumens and asphalts tend to stiffen with age.

The most appropriate type of waterproofing membrane will depend on the overall roughness and general state of the surface. Considerations will include:

- The size of steps between beams/slabs.

> A bonded membrane may not be suitable for use on a roof that has numerous irregular steps as it would be difficult to lay the membrane to avoid an unbonded channel forming along the steps. Chamfered fillets should be built into the steps or the roof screeded and then waterproofed. The same considerations should be applied before using a liquid membrane, and flexible materials for this purpose are often provided as part of the waterproofing system. Alternatively, a layer of clean aggregate or screed could be placed and an unbonded membrane laid on top.

- The surface condition of the concrete, i.e. weathered, dusty, porous, uneven, with sharp aggregate, etc.

- Overall shape of the roof, e.g. barrel vaulted, domed, flat, conical, with many upstands. This will be of importance when considering the use of a sheet membrane.

- Edge details.

- Provisions for drainage.

- Insulation.

> Consideration should be given to whether the insulation is to be placed on top of the roof (see also Section B4.1). For instance, it may be preferable to relieve the roof of dead load for structural reasons or for ease of maintenance, in which case a solar reflective bonded or liquid membrane (ultraviolet resistant) may be appropriate. Alternatively, the soil could be replaced by aggregate, which should be no-fines, washed and single sized to minimise vegetation growth. Should an unbonded membrane be used then an appropriate weight of insulation will be necessary as ballast to prevent the membrane from being lifted by the wind. hydrophilic membranes require the insulation layer to retain water in order for the membrane to continue to function effectively.

- Weather.

> Most bonded and liquid membranes will not bond to wet concrete; some systems have a tolerance to damp conditions if correctly applied.
>
> When planning roof waterproofing, consideration should be given to the effects of solar radiation on the roof. Possible solutions to this problem include temporary reflective coatings or the use of a tent.

*(a) Bonded preformed membranes*

Bonded membranes may be used on the walls of new reservoirs, but outward leakage may force the membrane off an external wall. Their use is usually restricted to waterproofing roofs.

Bonded waterproofing membranes are generally supplied in preformed factory manufactured sheet form, comprising an adhesive layer and a polymer backing protection sheet. They are produced as rolls or as boards which are bonded to the substrate to form a continuous membrane. They can either be laid cold or heated as they are unrolled, and usually require a substrate primer.

> When selecting a bonded membrane system, consideration should be given to whether the reservoir is to remain in service during application and the possibility of primer penetrating the roof, contaminating the water supply.

They are usually liquid and vapour-proof and relatively flexible, provided that the lap joints between sheets are reliable. However, certain types of sheet are prone to creasing or rippling, and the adhesive may not be sufficient to maintain continuity at the joints, leading to leaks.

> A non-destructive technique is now available for locating leaks in bonded membranes, by tracing electrical paths through the waterproofing layer, which on non-conductive membranes only exist at defects. Care must be taken to avoid damage to the membrane during the test, and atmospheric conditions must be considered during interpretation of the test results.

Membranes are manufactured in various grades and the required grade should be carefully selected for each individual application: i.e. a thin grade with protection boards may be more economic than a heavy duty unprotected membrane. The grade selected will ultimately depend on the construction traffic expected, on the type of backfill and the way it is placed.

When applied directly onto the concrete roof, sodium bentonite sandwich-type membranes are applied as unbonded, but will become bonded after wetting. Further details are given in Section A4.2.3 (c).

> Additional points to be noted on the application of bonded preformed membranes are as follows.
>
> - Avoid sharp corners. Infill corners with fillets and form chamfers wherever possible.
>
> - The sealing of leaking joints and cracks prior to application of the membrane is not necessary but is considered to be a prudent precaution, particularly against structural deterioration.
>
> - The concrete surface should usually be dry before application of the primer, although some systems have a tolerance to damp conditions.
>
> - Surface preparation and priming of the concrete are important with this type of membrane to achieve an effective bond. The membrane may require a water-jetted, grit-blasted or vacuum-ground surface.
>
> - Primers must be dry before laying the membrane, to avoid blistering.
>
> - Adhesion difficulties may be encountered if the concrete surface has been rendered.
>
> - Membranes with adhesive on both sides should be avoided as they are difficult to lay.
>
> - Membranes should be prevented from bonding over joints, including the roof/wall joint, to accommodate movement.
>
> - Laps must be properly rolled.
>
> - Roof laps must be laid so as not to trap water; i.e. each membrane sheet should be laid on the upslope side of the previously laid sheet.
>
> - Watertightness at laps may be difficult to achieve with dimpled membranes.
>
> - Organic sealants or polymer modified mortar/cement are more efficient upstand rebate sealants than cement mortar, which could shrink.
>
> - It may be difficult to undertake water testing on a sloping roof because considerable depths of water may be required even if small areas are tested. Spray testing is recommended in these situations.
>
> - Birds like to roost on exposed or reflective membranes and may cause damage by pecking at the surface.

*(b) Unbonded preformed membranes*

Unbonded membranes take the form of either flexible waterproof sheets with welded joints, or mats connected together by a suitable overlap to achieve a continuous water- and vapour-resistant barrier.

Unbonded membranes rely on perfect joints between sheets for their watertightness. The sealing of leaking joints and cracks prior to laying the membrane is not necessary but is considered to be a prudent measure.

> Watertightness of joints in the membrane can be tested by an air test, in which double joints are formed in the sheeting and air pressure is developed between them, or a vacuum test, using a vacuum box and soap solution. One manufacturer has developed a copper wire device for checking the welds between sheets by means of a spark test.

Unbonded membranes may be laid within the insulation layer (provided it is clean and free from contamination) or over existing membranes. However, where the membrane is laid directly onto a concrete roof, sand or a geofabric protective layer should be provided (underneath a relatively thin unbonded membrane) to prevent puncturing by pedestrians and/or light plant used during construction work. The erection of signs on site to warn workers that an unbonded membrane system has been installed is considered a prudent measure during construction work. Should the membrane be punctured after laying, and a leak result, the water is free to travel over the whole roof (beneath the membrane) and the source of the leak is virtually undetectable. This is also true for animal or vegetable penetration after laying.

This problem is not likely to occur for hydrated sodium bentonite membranes, as punctures effectively self-heal (see Section A4.2.3 (c)). Unbonded membranes tend to be manufactured in wider rolls than bonded membranes and hence will have fewer site joints for the same area; however, this may require the use of larger plant to lift them onto the roof and the heavier rolls could generate localised high stresses in the structure. Unbonded membranes will accommodate roof movement.

Unbonded membranes are unsuitable for walls. As a roofing material they are susceptible to vandalism.

> As for bonded membranes, the sheeting could be carried past the edge of a roof and down the wall to waterproof the roof/wall joint or to form a lining to a French drain (see Section B2.1).

*(c) Unbonded hydrophilic membranes*
Hydrophilic membranes are composite materials consisting of a carrier material (often sacrificial) of geofabric, paper, high density polyethylene (HDPE) or isoprene rubber, which are filled or impregnated with dehydrated natural sodium bentonite, which is actively hydrophilic.

The sodium bentonite reacts with water, resulting in swelling and the production of an impermeable gel phase, which seals the surface to which it has been applied. Such membranes are laid unbonded, but on hydration generally become bonded to the surface. They must however, be adequately confined by the insulation layer to prevent excessive swelling.

The rate of hydration of the hydrophilic phase is dependent on the conditions of the site, i.e. the amount of water available for hydration and the water resistance of the carrier material. Geofabrics have little resistance to water and therefore react rapidly, while HDPE has a high resistance to water and hydration will only occur initially at joints and at punctures or fixing points in the carrier, which will self-heal on hydration. Care should be taken with the fast-acting materials to ensure that hydration does not occur prior to fixing. Where long periods of site exposure are unavoidable, sheets with increased weather (water) resistance should be used. If a sheet becomes hydrated and swells prior to covering it should be replaced.

Surface preparation does not need to be as rigorous for these types of membrane, as surface irregularities of up to ±10 mm can be accommodated on hydration. Joints between sheets are constructed by self-sealing overlaps. Bentonitic sealants are also available for added protection.

> There are a number of important points to note on the use of these types of membrane.
>
> - Care should be taken with some types of sheet to ensure that the granular sodium bentonite is not lost when they are cut or nailed to the substrate.
>
> - Unlike other membrane systems, for hydrophilic membranes to function effectively it is essential that they be kept wet at all times, as the gel will only reconstitute slowly if it is allowed to dry out which may result in leakage during the next wet period. The insulation layer may need to be modified by the inclusion of a minimum 100 mm thick layer of 12 mm single sized washed gravel and a geotextile membrane to ensure sufficient water is retained.
>
> - The hydration of sodium bentonite may be inhibited in the presence of moderate concentrations of salts, particularly calcium, chloride and trivalent ions.
>
> - Sodium bentonite membranes are also susceptible to root damage.

*(d) Liquid membranes*

A liquid membrane is a one- or two-component moisture or chemically curing solution which is applied to the surface by either spray, squeegee or brush and dries to form a waterproof membrane of up to several millimetres thick. The principal advantages of this form of material are that it can effectively be applied to uneven or slightly stepped surfaces and site repairs can often be undertaken using the same material.

> Care should be taken with all these systems during application to 'live' reservoirs to ensure that the liquid does not penetrate the roof and contaminate the water supply, as the uncured materials may not satisfy the requirements of the Secretaries of State's or Water Byelaws scheme approval (see Section C1.3.1).

Several liquid systems give the option of fibre reinforcement either with separate fibres or with fabrics. The fabrics are used between coats of the material and provide additional tensile strength over cracks and steps.

> Some points to be noted areas follows.
>
> - Liquid membranes should not be used on a surface that is dusty or friable, as adhesion will be lost. Surfaces containing sharp aggregate should first be covered with a screed. Water-jetting, grinding, vacuum-assisted grinding or shot blasting may be used to obtain a dust-free surface.
>
> - Certain liquid membranes have the ability to remain bonded to the concrete against the upwards movement of water vapour. Loss of bond due to water vapour pressure can be a more significant problem in lightweight concrete, e.g. lightweight concrete screed.
>
> - The availability of application equipment may be limited and thus prices may be non-competitive.
>
> - Membranes may need to be de-bonded over joints to accommodate movement; advice from the manufacturer should be sought.
>
> - Liquid membranes also tend to be sensitive to the weather conditions at the time of application and to workmanship.

### A4.2.4 Internal waterproofing treatments

*Walls and floors*
Among the thin surface coating materials used for this purpose are:

- bitumen emulsions and solutions
- epoxide resins
- elastomeric polysulphide
- polyurethane resins
- polymer modified cementitious (coatings, mortars or renders)
- reinforced sprayed concrete (reinforcement can be pinned to the structure and curing must be sufficient to prevent cracking).

These membranes/renders are easy to apply. Some of them can be applied in damp, but not wet conditions. They will not readily bridge cracks or surface blowholes, which should be filled before application of the material. Thin layers are in general fragile and may require protection against wear.

The traditional material for internal application is waterproof cementitious render, which has been used in situations with up to a 30 m head of water. The render is a stiff sand and cement mortar containing one or more admixtures, and is applied in three or more continuous coats to a surface that has been prepared by scabbling or grit blasting. The mortar is usually of the order of 3:1 sand:cement and the admixtures may be hydrophilic materials (in the presence of water these seal and close the pores in the mortar) or hydrophobic materials (these inhibit the wetting of the pores). Unlike other systems, waterproofing renders can be applied in wet conditions and even against a considerable head of water. They should be applied by specialist contractors.

The leaching of metals from cementitious materials is potentially a problem in all areas, including new construction. The dwi have sponsored and published an international review of the composition of materials likely to be used in cementitious systems in water retaining structures[15].

*Roofs*
When a roof is leaking, waterproofing the top surface is always the best solution. This may, in some cases, not be possible, e.g. when access is restricted. In other cases it may be uneconomic.

In such cases, the provision of a waterproofing treatment on the inside may be the only solution. However, sealing of significant cracks would be necessary prior to application and hydrostatic pressure testing after completion of the work.

> Weight restrictions placed on a roof as a result of structural assessment may make any external construction activity impossible. Where this is the case, care must be taken when carrying out surface preparation prior to internal waterproofing, e.g. cutting out cracks for repair, to prevent stresses being induced which are sufficient to further damage the roof.

Generic groups of internal waterproofing treatments suitable for roofs are:

- polymer modified cementitious (mortars/renders)
- epoxide resin (which can be formulated to give a range of properties)
- reinforced sprayed concrete (reinforcement can be pinned to the structure and curing must be sufficient to prevent cracking).

As with materials for walls and floors, a period of forced ventilation may be necessary to remove all traces of solvent prior to the reservoir being returned to service.

### A4.2.5 Surface coatings

> For the purposes of this report, surface coatings are differentiated from waterproofing membranes/treatments in that they are not primarily used for waterproofing, but to provide a surface to meet other water quality or durability requirements e.g. to prevent bacterial growth or resistance to deterioration of a concrete or masonry substrate. They may, however, provide some degree of waterproofing.

*External surfaces*

A number of coatings are available that would be suitable for application to the external surface of a roof. The materials act to reduce the permeability of the concrete (especially a poor one) and, to some extent, seal hairline cracks. Surface coatings cannot be relied upon to act as a waterproofing membrane unless they are applied in conjunction with good drainage to shed all surface water.

Generic groups are:

- modified cementitious
- bitumens
- polysulphide
- resins
- silicones, silanes and siloxanes.

*Internal surfaces*

The provision of a surface coating on the inside of a reservoir would have three purposes. First, a coating can provide resistance to the ingress of carbonation (particularly on roof soffits) and harmful salts into concrete, and thus provide additional protection to reinforcement. Second, a coating can provide a degree of waterproofing to porous or finely cracked concrete or brickwork. Third, a coating can present a dense and smooth surface that reduces the risk of dirt or bacterial deposits, to which the relatively rough surface texture of even a high quality concrete is susceptible, and also considerably ease cleaning operations.

The principal limitation of a surface coating is that it will generally not seal large existing cracks, and only flexible materials will span cracks occurring after application. Generic groups are:

- polymer modified cementitious
- epoxide resins
- elastomeric polysulphide.

> Points to note are:
>
> Coatings alone are not satisfactory where the concrete contains blowholes. If a coating is to be used the blowholes should be made good, possibly by the use of a fairing coat, prior to application. A polymer modified cementitious mortar with a cement/polymer emulsion bond coat would be more appropriate. This is also likely to be a more durable solution.
>
> Bituminous compounds and some brush applied epoxies have been used in the past but they have not proved to be durable and can impart a phenolic taste to the water.
>
> Some surface coatings will generally not tolerate acidic conditions, and require the water to have a minimum pH of 7.2 and a minimum total hardness of 140 mg/l. However, the pH of water leaving treatment works (usually pH 7 or 8) may change on the way to service reservoirs for a number of reasons including air entrainment, reduction in the residual chlorine content, leaching of lime from cementitious materials, etc.
>
> Some cementitious based coatings may be susceptible to attack by aggressive water. If so, they would only provided a limited period of sacrificial protection to concrete or mortar.
>
> Coatings should be selected to be chemically resistant to the water that is to be stored in the reservoir and the advice of the manufacturer should be sought if there is any doubt.

### A4.2.6 Metalwork paint systems

Three paint layers were traditionally used: a priming layer in contact with the steel, designed to adhere well to the metal and which may contain particles of zinc, zinc chromate or zinc phosphate, which are hydrophilic corrosion inhibitors; an undercoat layer (on top of the primer) to provide coating thickness; and then the finish layer to give colour and texture.

> There are a number of points to note on the use of zinc-based priming coats.
>
> - Zinc chromate is water soluble and complete coverage will be required to ensure contamination does not occur.
>
> - Zinc phosphate is now widely accepted as a corrosion inhibitor.
>
> - Pure zinc acts to give sacrificial protection and specific guidelines on preparation, zinc content, etc., should be followed (see also Section A4.2.1 (a)).

Modern high build paints combine the properties of the intermediate and top coats to save application time, costs and the possibility of intercoat contamination.

All paints are permeable to air and water vapour to some degree and for a particular material the protection afforded is directly proportional to thickness.

Generic groups include:

- bitumen
- drying oils
- chlorinated rubbers
- vinyl and vinyl esters
- acrylic resins
- epoxide and polyester resins
- polyurethane
- polymer modified cementitious.

Bitumen is not a good medium and is difficult to apply. However, it is cheap and can be used in thicknesses up to several millimetres. It is impermeable to water and may be suitable for external use. Some bituminous compounds can impart a phenolic taste to water and are therefore not acceptable for internal use.

Drying oils include traditional paints based on linseed and tung oils. Modern paints of this type are based on alkyd resins. They are low cost but not suitable for underwater applications.

Non-converting resins, e.g. chlorinated rubbers (now being phased out for COSHH reasons), vinyl copolymers and acrylics dry by solvent evaporation. They are generally expensive (economics will depend on the number of applications required to achieve the required film thickness) but give good protection in aggressive environments. Their advantage lies in maintenance as old weathered films can be simply overcoated after removal of loose contaminants.

Epoxy resins and polyurethane paints have excellent chemical resistance and durability. High film thicknesses may be applied in single coats, and they can be formulated with fast drying solvents, enabling work to be completed with minimum 'downtime'. Polyurethanes are less degradable than epoxies but are more expensive.

> Paint systems that dry/cure by means of solvent evaporation have the added risk of solvent retention (possibly leading to contamination problems) and additional project costs due to increased ventilation requirements and extended 'downtime'. Where possible, solvent free products are recommended to eliminate these problems.

Two-component, thixotropic, polymer modified cementitious coatings can be formulated to be highly resistant to moisture and chloride, and because of their high alkalinity, they passivate the steel surface. They have some degree of elasticity and will accommodate some movement.

> It is recommended that, where possible, moisture tolerant paint systems should be used, as achieving a 'dry' substrate cannot be guaranteed in the reservoir.

### A4.2.7 Crack injection

Crack injection should only be contemplated when it has been proven by investigation that a crack is not moving through structural loading or thermal movement. This is likely to be the case for newly constructed covered reservoirs once they are in operation (i.e. after approximately 6 months). Some cracks in new structures will tend to close as the concrete absorbs moisture and expands, or autogenous healing may occur, particularly where concrete containing pozzolanic cement replacement materials (ggbs or pfa) have been used.

When sealing leaks, care should be taken to find the source and to seal close to it. A leak may appear only after water has travelled many metres laterally. This makes it difficult to find the source, but sealing at the exit only forces the flow to the next path of least resistance.

In a few instances, crack filling is of structural significance, and crack injection under pressure, using a suitable material, would be an appropriate technique.

Crack injection can be used to seal cracks with a width down to 0.1 mm. It is unlikely that any crack in a reservoir will be dry, and therefore a moisture tolerant material should be used. Where non-structural crack sealing is required, a latex emulsion can be used which hardens by dispersion of water into the concrete, depositing a rubbery mass, which is water resistant (though not under hydrostatic pressure) and able to accommodate movement. Such a material can be applied either by gravity or by force-feeding against gravity from the lowest part of the crack.

Chemical grouts are frequently selected for sealing leaks through cracks. They are generally fluids that can be pumped into a flow channel at pressures similar to those needed to cause water flow through the same channel. After a time, the grout undergoes a phase change to gel or elastic foam. Two types are available: for wet fissures, these are the two-part, minimally or non-foaming, highly flexible polyurethane resins, mixed at the point of injection; for sealing stronger water ingress or egress, these resins are not suitable because they do not set sufficiently rapidly and consequently are flushed out. In such cases single-component resins may be used. The hardening of these resins is based entirely on their reaction with water; on contact with water the development of a large volume of foam takes place within seconds or several minutes (according to the type used) which can stop water ingress. Owing to their structure they cannot ensure permanent sealing. However, two-part and single-component resin systems are complementary, when used in combination.

The selection of injection systems will depend upon the nature and size of the cracks, the temperature of the structure, the time allowed for repair before loading the structure and the injection method proposed. Generic groups of materials include:

- epoxide resin
- polyester resin
- polyurethane resin
- Styrene Butadiene Rubber (SBR) and acrylic latex emulsions
- polymer modified cementitious grout.

Care must be exercised with acrylic gels which are water activated. They are composed of unlinked monomer units, which when unconfined will expand continuously. Problems have been encountered where joint movement has cracked the gel, exposing fresh surfaces to water, and has resulted in excessive expansion through absorption.

Polyurethanes are preferable to acrylics as they are unable to expand further, after setting. The single-component systems are best used as a temporary measure; they react quickly in the presence of water and prevent leakage.

Care should be taken when grouting larger areas to ensure that any drainage systems including wall backing layers are not blocked.

Most two-part injection systems are highly toxic and/or corrosive in the uncured state. Acrylate and acrylamide grouts, which were banned in the UK approximately 10 years ago, are returning to the market in a modified form.

### A4.2.8 Crack fillers and sealants

Crack sealing with a non-flexible sealant, as for crack injection systems, should only be contemplated where it has been proven by investigation that a crack is no longer moving. In addition, it must also be assumed that thermal movement will take place in all cracks upon the removal of insulation.

*Non-flexible sealants*
Generic groups of materials suitable are:

- epoxy putty (some types are designed to accommodate movement)
- polymer modified mortar (or grout for cracks up to 1.0 mm)
- polymer resin (generally suitable for cracks up to 1.0 mm).

*Flexible sealants*
Generic groups of materials suitable are:

- thermoplastics (either applied hot or cold)
- chemically curing elastomeric compounds, e.g. polysulphides, polyurethanes, silicones and polymer modified epoxide (these accommodate far greater movement and are more durable)
- polymer modified cementitious slurries with fibreglass reinforcing tape (these accommodate some movement).

Primers should be selected to be compatible with the sealant used and the substrate, and to be suitable for the service environment.

> Some epoxy primers can be used on damp substrates to provide a suitable surface for sealant application.
>
> Sealants having a low polymer content and a high filler content have been known to break down after only a few years' exposure. In general, a higher polymer content gives better performance characteristics. Good quality sealants should be specified that comply with a relevant British Standard, where one exists.

*Preformed sealants*
Generic groups of suitable materials include:

- hypalon – overlay strip
- chlorinated polyethylene – overlay strip
- neoprene – insertion strip
- EPDM – insertion or overlay strip
- bitumen rubber
- butyl rubber.

> The adhesive at the edge of strips has been reported to fail after a relatively short period (approximately one year) in conditions of continuous immersion.
>
> When using an insertion strip, the cost of cutting the groove should also be taken into account.
>
> Bonding between strips may be difficult where two strips must be bonded together to form a continuous strip or to accommodate intersecting cracks.

*Crack filling*
If the need for structural continuity is not a requirement, then static cracks can be sealed against moisture by brushing cement or neat cement grout into them. Another method is to seal cracks using a low viscosity latex emulsion, which provides protection against water penetration by allowing a latex deposit to build up within the crack. However, crack filling in this manner will only accommodate slight movement (approximately 10% of the crack width). Wider cracks may be filled with a latex-cement mixture to which a thickener may be added to promote gelling and help retain the material inside the crack, or by the use of a crack injection system.

### A4.2.9 Joint fillers and sealants

*Joint fillers*
Joint fillers, usually manufactured from polyethylene, are supplied as preformed sheets, which are cut to size on site (some products are available pre-cut to size) and are used to provide support to a sealant under hydrostatic pressure.

It is important to select the joint filler with regard to its suitability in terms of compression load, sealant compatibility ease of installation and, where required, resistance to hydrostatic loading.

*Joint sealants*
Generic groups of suitable materials are:

- thermoplastics, e.g. bitumen rubber, applied hot or cold
- chemically curing elastomeric compounds, e.g. elastomeric polysulphide or polyurethane.

In general the rate of cure and movement accommodation factor of the different generic types of joint sealants are set out in Table A4.1.

**Table A4.1** Relationship between rate of cure and movement accommodation factor for generic groups of sealants

| Sealant type | Rate of cure* | Movement accommodation |
|---|---|---|
| Thermoplastic | Very fast | Low/medium |
| Elastomeric (2-part) | Fast | High |
| Elastomeric (1-part) | Slow | High |

* The faster the rate of cure of the sealant, the better it will be at accommodating early movement and early immersion in water.

The level of cure achieved before the reservoir is filled should be sufficient to guarantee that water quality will not be compromised. One-part sealants are generally not considered appropriate for permanent immersion in water.

Further details on joint sealants can be found in CIRIA publications TN128 (1987)[16], TN144 (1992)[17], SP80 (1992)[18], R128 (1992)[19] and FR/CP/17 (1994)[20].

Substrate primers may be necessary in some circumstances. These should be selected to be compatible with the sealant used. The rate of cure of some primers is temperature sensitive.

Careful sealant specification should eliminate problems of age hardening and the use of materials that will support bacterial growth.

> Most sealants are difficult to apply to joints in walls and floors where the concrete has been saturated, as they require a dry surface to achieve a good bond. Torching will only dry the concrete surface, and moisture migrating from the concrete core can still cause the sealant to fail. Possible solutions include using a polythene tent to create a dry atmosphere or to using a damp tolerant sealant system. Care must be taken to ensure that back pressure does not remove the primer film before adhesion occurs.

*Preformed strip sealants*
The use of preformed strip sealants with an unbonded section over the joint can provide a more reliable method of joint sealing than the use of a sealant material. Suitable generic groups are as for preformed crack sealants.

### A4.2.10 Waterstops

BS8007: 1987 states *'it is not necessary to incorporate waterstops in properly constructed construction joints'*; it is now, however, considered a prudent measure to include a waterstop in all types of joint.

Generic types include:

- barrier (passive) waterstops, e.g.
    - bitumen plugs
    - metal waterstops
    - rubber waterstops
    - plastic waterstops (generally PVC)
- hydrophilic waterstops
- co-extruded waterstops
- post-grouted tubes

Cavities are sometimes provided in the centre of joints for subsequent filling with a bitumen plug. However, it is impossible to make a satisfactory compromise between the requirement for fluidity and the need that the material should not contract on setting, flow out of the joint or become brittle.

The concept of a rigid metal waterstop is now only used for kicker joints where full height wall pours are to take place. In many older structures they were used in horizontal construction joints exposed to full hydrostatic pressure. Metal waterstops were generally steel strips, which were difficult to join and susceptible to corrosion, at poorly formed joints, unless bitumen coated. Copper waterstops are not intended for use in service reservoirs.

Rubber and PVC waterstops are generally sections that are wholly or partially embedded in the concrete during construction so as to span the joint and provide a permanent watertight seal.

Rubber waterstops are used primarily where ground movements are anticipated. Difficulties in jointing and forming junctions on site may be experienced as the material has to be hot vulcanised to form an effective joint. Equipment and methods are, however, available to make reliable and effective site joints. 'Low modulus' PVC waterstops have been accepted in some applications as they offer comparable performance with simplified fabrication and assembly, though where cyclic or large magnitude movements are expected rubber waterstops are still preferred.

PVC waterstops are the most commonly employed product in modern structures. They can be formed into complex shapes more easily than rubber waterstops.

Hydrophilic waterstops rely upon the sealing pressure developed by swelling (due to water absorption by a hydrophilic polymer) when they are adequately confined within the concrete, and may provide additional security. They may consist of a hydrophilic polymer, compounded hydrophilic polymer and rubber in strip form. Some types are suited to repair applications as they do not require breaking out, slot cutting or the reinstatement required by conventional waterstops. Care should be taken with these materials to prevent premature expansion of the hydrophilic element (any affected sections should be cut out and replaced) or where high concentrations of salts are expected (which may reduce the swelling capacity). See also Section A4.2.3 (c), which contains additional information on the behaviour of hydrophilic membranes.

Co-extruded waterstops are similar to barrier waterstops except that the bulb-ends are co-extruded with a hydrophilic material designed to perform as described above, providing additional contact pressure at the bulb-end should any water reach that point.

Tubes incorporated into the joint during construction can either be grouted as soon as the concrete has hardened or after testing the joint for water-tightness. They are generally considered as a secondary system and may be used in association with a conventional waterstop to give a degree of additional protection or for ease of remedial works.

> Mistakes made in joint design and waterstop installation may be difficult and expensive to repair at a later stage. Voids can be formed in the concrete adjacent to the waterstop if it is not fixed adequately when placing the concrete or adequate compaction is not given to the fresh concrete at the waterstop.
>
> Further guidance on the use of waterstops is given in CIRIA publication FR/CP/17[20].

### A4.2.11 Large volume grouting

Large volume grouting is an established technology in mining and tunnelling, which is now gaining wider recognition. Leakage from reservoirs may result in soil erosion, which develops as water flow increases. The resultant voids are suited to treatment by large volume grouting techniques. Grouting may be used for a number of purposes including:

- void filling
- ground strengthening
- reducing ground permeability
- leak sealing.

However, individual grout formulations will not be suitable for all purposes. Grouting the ground beneath a reservoir might decrease the rate of leakage to be within acceptable limits; it cannot be guaranteed to stop it.

*Cement based grouts*
Cement based grouts are generally rigid and should only be used when further movement is not anticipated. Cement/bentonite grouts may retain some flexibility for long periods of time.

Cement based grouts are both cheap and effective. Unfortunately they will not penetrate ground with low permeability because the cement particles are too large to pass through the narrow throat passages.

*Chemical grouts*
Chemical grouts can be used for lower permeability ground, but they have many disadvantages such as susceptibility to leaching, low strength (e.g. silicate gels), toxicity, etc. They would not be used for void filling.

> CIRIA Report No. 95[21] deals with the problems of chemical grouts, highlighted environmental pollution caused by acrylic and acrylamide grouts. The report lists the ways ground treatment materials can enter groundwater supplies. These include spillage or malfunctions of the injection system, over-dilution of chemicals such that setting does not occur, and degradation of the set grout *in situ*.
>
> The Health and Safety Commission's Control of Substances Hazardous to Health (COSHH) Regulations (October 1989)[22] make it a legal requirement that suppliers detail the chemical content and the dangers involved in using the product.

The two-component polyurethane systems are widely used for ground consolidation and water stop. Properties range from hard, stiff solids to flexible soft foams.

With the rapidly reacting two-component resins the isocyanate component reacts with both the water and the polyol. When penetrating into the wet strata the mixture comes into contact with water, incorporates it, and strong foaming takes place. In the course of the injection, this foam is displaced by the continuous resin feed. The newly fed resin encounters less water and consequently foams to a lesser extent and this becomes increasingly more dense and more solid, up to a point where the resin does not encounter further water and sets to form a compact polyurethane without foaming. The result is a 300-1000 mm thick, solid, water-tight jacket, surrounded by shells of elastic polyurethane which become less dense the deeper they extend into the strata.

> Problems have occurred where grouts have degraded and leached into aquifers. Acrylamide replacements, often marketed as environmentally friendly or non-toxic, are in fact merely less toxic than others.
>
> 'Alternative grouts' cause problems; silicate based grouts (some banned in Austria and Japan) are commonly used in the UK; formaldehyde and isocyanate grouts are especially poisonous and must be avoided.

## A4.3 MATERIALS SELECTION

The procedures that should be followed to identify the most suitable materials category for waterproofing and repair are set out in Section A3. Sections A4.1 and A4.2 give details of the available approval schemes and information on the generic properties of the materials within each category. Selection of an individual material for a particular application will require consideration of a number of factors.

This Section, by means of a flow chart, outlines the procedures that should be followed in order to select the most appropriate material.

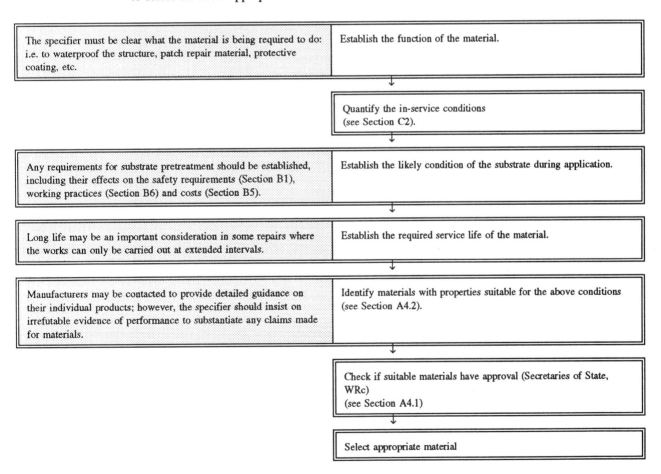

# Appendix AP I Causes of defects in reservoirs

## AP I.1 CORROSION OF METALWORK (INCLUDING STRUCTURAL METALWORK AND ACCESS LADDERS, ETC.)

A number of different types of metal have been used structurally in the construction of reservoirs. These include:

- steel (carbon and stainless)
- cast iron
- wrought iron
- aluminium
- galvanised steel.

Many of them have also been used for fittings within reservoirs. The principal causes of corrosion of many of these metals, detailed in Sections A2.4.1 and 2.4.7, are:

- exposure to the moist environment above water level within the reservoir
- exposure to the stored water
- exposure of prestressing bands to water leaking outwards or groundwater
- bimetallic corrosion.

The deterioration of most metals used in construction (and susceptible to corrosion) can be prevented by the use of a protective system, e.g. by painting or by coating beams with a layer of concrete. Corrosion will only then occur if the protective system used is inadequate or defective (through either poor specification, poor application or inadequate maintenance).

Metals in contact with water corrode because of their thermodynamic instability. Water in reservoirs contains dissolved solids and gases, which may affect its corrosive properties in relation to the metals with which it is in contact. Condensed water (condensate) on the surface of metalwork will have a different composition from that being stored in the reservoir, and hence its corrosion behaviour will be different.

### AP I.1.1 Ferrous metals (cast and wrought iron and steel)

*Exposure above water level*
Ferrous metals (steel, cast and wrought iron) in dry air are protected by a thin oxide film which forms on the surface of the metal. In the presence of electrolytes, however, protection is lost and the metal will corrode to form a wide range of iron oxides and hydroxides, the final product of corrosion being the familiar reddish brown rust ($Fe_2O_3.nH_2O$).

Once the relative humidity rises above a critical, but comparatively low, value (approximately 70–80%), the rate of rusting in the air space (where condensation may occur) is dominated by minute concentrations of impurities such as sulphur dioxide derived from the atmosphere.

The corrosion resistance of stainless steels in all reservoir environments is significantly enhanced by the inclusion of chromium and nickel as alloying elements. These stabilise the passive film and prevent its breakdown.

> The use of grade 316S31 stainless steel should be encouraged for all internal metalwork (particularly ladders, etc.) in order to minimise maintenance requirements.

*Exposure below water level*

Corrosion in water is a complicated phenomenon, which is beyond the scope of this document. The main considerations involved are the composition and surface condition of the metal, the quality of the water (particularly the chloride concentration) and the operating conditions, particularly pH. Low pH environments increase the rate of corrosion.

### AP I.1.2 Aluminium

*Exposure above water level*

Aluminium alloys as a group generally corrode in air to form grey aluminium oxide, which is protective in most atmospheric environments.

*Exposure below water level*

Immersed aluminium and its alloys also have good resistance to attack by pure condensate water. Their behaviour when immersed in potable water is dependent on the dissolved solid content. Of the major British Standard alloys only those that contain copper as a major alloying constituent are likely to corrode in chloride solutions. The combination of carbonate, copper and chloride ions in water is, however, aggressive and as little as 0.2 mg/l copper in hard waters can initiate pitting. This combination of ions required to initiate corrosion indicates that some waters will be more aggressive than others. However, in waters of all types the rate of pitting falls off rapidly with time.

> The use of aluminium is not generally recommended within reservoirs owing to its susceptibility to corrosion and possible water quality problems.

### AP I.1.3 Galvanised steel

The zinc coating (provided by galvanizing) is more anodic than the steel substrate and so will corrode preferentially (see Section AP I.1.4) protecting the steel. Zinc, however, is susceptible to rapid attack where high levels of residual chlorine are present.

> The additional short period of protection provided by the coating may not justify the increased cost of this material.

### AP I.1.4 Bimetallic corrosion

When a metal is immersed in an electrolyte the chemical reactions resulting in the movement of electrons (which occur at the surface) establish an electrical potential. This potential is different for each type of metal and may be different for different compositions (or structures) of the same metal (e.g. welds, particularly on ladders). The ranking of metals in order of their electrode potential is known as the galvanic series (see Table AP I.1). Where two metals are electrically connected and bridged by an electrolyte, a corrosion (galvanic) current will flow from the more electro-positive metal (least noble, or anode) to the more electro-negative metal (most noble, or cathode) (N.B. Corrosion current flows in the opposite direction to electron movement and should not be confused with it).

When the potential difference between the two metals is sufficiently large, the current flow can affect the rate of corrosion of the metals, by accelerating the rate of corrosion of the anode and suppressing the rate of corrosion of the cathode. The magnitude of the effect being dependent on the potential difference established.

There are four requirements for bimetallic corrosion to occur.

1. An electrolyte must bridge the two metals.
2. The two metals must be connected electrically.
3. The difference in the potentials of the two metals must be large enough for a significant galvanic current to flow.
4. The cathodic reaction of the more noble metal must continue unobstructed.

If any one of these requirements is not net bimetallic corrosion will not occur. There are a number of features that affect the corrosion rate. These include: the conductivity of the electrolyte, the relative area of the metals, degree of aeration and flow rate of the electrolyte.

**Table AP I.1** Galvanic series of metals and alloys (abbreviated) in a strong electrolyte

| | | |
|---|---|---|
| Zinc | | At greatest corrosion risk (anodic or least noble) |
| Aluminium 2S | | |
| Aluminium 245-T | | |
| Steel or iron | | |
| Cast iron | | |
| 13% Cr-iron (active) | ⎫ | |
| 18% Cr-8% Ni steel (active) | ⎬ Stainless steels | |
| 18% Cr-8% Ni-3% Mo steel (active) | ⎭ | |
| Inconel (80%Ni, 13%Cr, 7% Fe)(active) | | |
| Hastelloy A (60%Ni, 20%Mo, 20%Fe) | | |
| Hastelloy B (65%Ni, 32%Mo, 1%Fe) | | |
| Brasses (Cu-Zn) | ⎫ | |
| Copper | ⎬ | |
| Bronzes (Cu-Sn) | ⎭ | |
| Inconel (passive) | | |
| 13% Cr-iron (passive) | ⎫ | |
| 18% Cr-8% Ni steel (passive) | ⎬ Stainless steels | At least corrosion risk (cathodic or most noble) |
| 18% Cr-8% Ni-3% Mo steel (passive) | ⎭ | |

## AP I.2 CONCRETE AND BRICKWORK DETERIORATION

The principal causes of deterioration are poor workmanship, poor design detail, chemical attack and physical attack.

### AP I.2.1 Poor workmanship

A number of defects that originate at the time of construction result from poor workmanship. In concrete construction the two most common deficiencies, which result in porous concrete, are air pockets and honeycombing. Air pockets or entrapped air are usually the result of insufficient compaction (vibration). Honeycombing occurs when the cement paste does not completely fill the spaces between the coarse aggregate particles. It may be caused either by incomplete compaction or by grout loss at joints in the formwork. Other defects that may occur as a result of poor working practices include:

- poor grouting of shutter tie-holes (walls), pipes and ducts[23]
- voids beneath tie-hole tubes (walls) and other box-outs/void formers
- poor preparation of kickers
- cold joints forming leakage paths
- misplaced or moved waterstops (see also Section AP I.5)
- loss of section due to failure of previous repairs
- use of filter bed gravels as aggregate, introducing chlorides, etc., into the concrete.

In masonry construction the most frequent problems are related to mortar composition and bond to bricks.

## AP I.2.2 Chemical attack

*(a) Sulphate attack*
Sulphate solutions react with the hydrated calcium aluminate phases and portlandite (Ca(OH)$_2$) in the cement paste. The products of these reactions, gypsum (CaSO$_4$.2H$_2$O) and calcium trisulphoaluminate (ettringite), occupy a considerably greater volume than the compounds they replace. Internal stress generated by the growth of the reaction products leads to disruption of the paste[24].

The reaction of sodium sulphate (the most common naturally occurring sulphate salt in groundwater) can be written as follows:

$$Ca(OH)_2 + Na_2SO_4 + 2H_2O \rightarrow \underset{\text{Gypsum}}{CaSO_4.2H_2O} + 2NaOH$$

In flowing water all the CaSO$_4$.2H$_2$O may be leached out.

The reaction with calcium aluminate hydrates can be generalised as follows:

$$2(3CaO.Al_2O_3.12H_2O) + 3Na_2SO_4 + 13H_2O \rightarrow \underset{\text{Ettringite}}{6CaO.Al_2O_3.3SO_3.31H_2O} + 2Al(OH)_3 + 6NaOH$$

The reactions result in an increase in the alkalinity of the pore fluids (NaOH is produced) with potentially serious effects if the aggregate used is alkali reactive (see Section AP I.2.2 (c)).

Concrete or mortar (in brickwork structures) attacked by sulphates has a characteristic whitish appearance and damage is usually concentrated at edges and corners. Progressive cracking and delamination (a separation along a plane parallel to the surface[25]) often reduces the concrete to a friable or soft state.

*(b) Carbonation*
Carbonation to depths greater than 5 mm has only been reported in a few reservoir roofs. This slow rate of carbonation in most of the structure is probably due to the almost complete saturation of the concrete or mortar, reducing the rate at which carbon dioxide can penetrate into the cementitious material. The mechanism of carbonation is briefly described below.

Carbon dioxide (CO$_2$) present in the atmosphere dissolves in the concrete pore fluids to produce carbonic acid, which reacts with the cement hydrates and calcium hydroxide to produce calcium carbonate (CaCO$_3$), silica gel, alumina and ferric oxide. Associated with this reaction is the almost complete loss of alkalinity of the cement paste. This has severe consequences for reinforcement (see also Section AP I.4) or for iron or steel with a concrete protective coating. Steel in concrete is protected by a passive oxide film stabilised by the high alkalinity (pH > 12.5) of the concrete. The passive film is destabilised at a pH < 11.5. When the carbonation front (a zone approximately 3 mm wide in which the alkalinity falls to pH≈9) reaches the level of the reinforcement, the stability of the protective oxide film is lost and corrosion can begin[26].

The rate of carbonation is dependent on the moisture content of the cementitious material and its porosity, and is approximately proportional to the square root of time multiplied by a constant, which is related to the quality of the concrete. The rate is greatest where gaseous CO$_2$ can penetrate the pore spaces and there is sufficient moisture to just dissolve all the CO$_2$. The greatest depths of carbonation occur where the relative humidity is in the range 60–70%. The strength of the concrete and its weight generally increase with carbonation; however, irreversible shrinkage may result in crazing of the concrete surface.

*(c) Alkali aggregate reaction (AAR)*
Three types of alkali aggregate reaction (AAR) are known. These are alkali silica reaction (ASR), alkali carbonate reaction (ACR) and alkali silicate reaction.

ASR has been reported in relatively few UK structures but these include some service reservoirs. It results from the chemical reaction between alkali hydroxides (in the concrete pore solutions) and certain types of silica (in the aggregate), which produces alkali silicate gel, which can absorb water and swell, generating internal stresses resulting in cracking and disruption of the aggregate and cement paste. In some cases pop-outs are formed where an isolated reactive particle occurs close to the surface of the concrete. Alkali carbonate reaction and alkali silicate reaction are rarely encountered in the UK.

For any of these reactions to occur three conditions need to be satisfied simultaneously.

1. A sufficiently strong alkaline pore solution must be present. Alkalis usually originate from impurities in the cement; however, migration of alkalis from other sources can also initiate or sustain the reaction (see Sections AP I.2.2 (a) and AP I.2.2 (d)).

2. A proportion of reactive aggregate lying within the sensitive range.

3. Sufficient moisture in the concrete. (The high relative humidities of UK ambient conditions are usually sufficient for reaction to occur in roofs. Below water level walls and floors are likely to be especially sensitive.)

If any of these conditions are not satisfied the reaction will not proceed. Only in a few cases has the magnitude of these effects been sufficient to cause structural damage. Cracking may, however, allow high rates of outward leakage or contamination of the water supply from groundwater ingress.

Expansion is usually not uniform through the volume of the concrete; it is greater in localised sub-volumes of the concrete, and it is this that causes the distinctive pattern of regular bifurcating cracks.

ASR is unlikely to be a problem in recent construction (post-1987) if the rules laid out in Concrete Society Technical Report No. 30[27] have been followed.

*(d) Chloride attack*
Chloride ions introduced into the concrete from groundwater contaminated aggregate (filter bed gravels or unwashed marine aggregates, see Section AP I.2.1) or from the water stored in the reservoir react with the calcium aluminate hydrate phases in the cement paste in a similar manner to sulphates (see Section AP I.2.2). The products of the reaction are not expansive, and hence disruption and cracking of the cement paste does not occur.

Chloride ions can diffuse through the pore solutions of concrete and initiate corrosion of steel used as reinforcement (see Section AP I.4) or iron.

If sufficient concentrations of chloride salts are allowed to build up in porous materials, then disruption may occur due to crystallisation pressure (see Section AP I.2.2 (f)).

*(e) Low compressive strength concrete*
Low strength in concrete may originate from a number of features, the most common of which are poor mix design (low cement content, high w/c ratio, etc.), weak aggregate (see Section AP I.2.3 (a)) or chemical conversion of the cement (e.g. high alumina cement (HAC) concrete).

The latter is a special case in which one of the cement hydration products converts to another form with a consequent loss of strength. The rate of conversion is temperature dependent and is generally slow at ambient temperatures, with full conversion occurring in approximately 20 years. The loss in strength can be as much as 50%, which has in some cases led to catastrophic failures in structures. An example of this type of problem occurred in a reservoir built in the early 1970s, which when investigated showed a 30% loss of strength.

Low strength is usually manifested by deformation of the structure and cracking in response to applied loads.

*(f) Salt crystallisation*
Concrete, mortar or porous bricks saturated with salt solutions, particularly chlorides and sulphates, can suffer from crystallisation pressure damage during periods of drying. As water evaporates from the pore solutions, they become increasingly concentrated until saturation is reached. Crystals will then begin to grow within the pore space of the material. The mechanism of deterioration is similar to that associated with freeze–thaw damage (see Section AP I.2.3 (b)). As the crystals grow, expansion is impeded and the resulting internal stresses disrupt the matrix of the material. Crystallisation pressures in excess of 60 N/mm$^2$ have been measured for sodium chloride crystals. Salt crystallisation can occur in any porous solid exposed to salt laden solutions and subjected to periods of wetting and drying.

The salts (chlorides or sulphates) can originate from the material used as thermal insulation, from bricks in brickwork structures or from the local groundwater.

*(g) Acid attack*
Acid attack is only likely to occur in below groundlevel walls and floors of reservoirs, which come into contact with flowing low pH water. Acidic waters may result from either organic acids (in 'peaty' areas) or inorganic acids (in mining areas) being dissolved in the groundwater. Their effect is to react with the alkaline compounds of the cement matrix of concrete or mortar, dissolving and removing them, weakening the cement paste and increasing its porosity. pH adjustment of the water prior to storage in service reservoirs is usually sufficient to prevent attack occurring internally. Where the groundwater is static the extent of reaction is unlikely to be significant.

*(h) Carbonate imbalance*
In natural groundwaters the concentration of sparingly soluble carbonate ions ($CO_3^{2-}$) and soluble bicarbonate ions ($HCO_3^-$) is in equilibrium. Where significant quantities of $CO_2$ are dissolved in groundwater this equilibrium condition cannot be attained and hence a carbonate imbalance occurs. To correct this imbalance calcium salts are taken into solution. Concrete and mortar provide a ready supply of calcium salts and hence are susceptible to attack. The effects are similar to those of acid attack with the cement paste being weakened and porosity increased. In extreme cases all the calcium may be removed from the concrete leaving only the siliceous and aluminate components of the paste, which can be removed by light abrasion (see Figure AP I.1). Carbonate imbalance will not be corrected by pH adjustment if calcium salts are not added, and attack may occur internally.

## AP I.2.3 Physical attack

*(a) Aggregate unsoundness*
Soundness is the term generally given to the ability of an aggregate to withstand large changes in volume (expansion or shrinkage) as a result of changes in conditions. The physical causes of volume changes are freezing and thawing, changes in temperature and alternate wetting and drying.

Unsound aggregate results in a variety of problems from local scaling or pop-outs to severe surface cracking and disintegration of concrete or mortar over a considerable depth, and thus can vary from cosmetic effects to being of structural concern. Susceptible aggregates include porous flint or chert, some shales, mining waste, limestones, with laminae of expansive clay, and other materials which contain clay minerals, particularly of the montmorillonite or illite group.

**Figure AP I.1** *Effect of carbonate imbalance on concrete within a reservoir*

*(b) Freeze–thaw*
Scaling caused in freeze–thaw is the flaking of the surface of the material, and can occur in concrete, mortar and bricks. It is caused by the repeated freezing and thawing of capillary water. Ice occupies a larger volume than water (approximately 9% greater) and hence sets up a bursting pressure, resulting in a series of fine cracks parallel to the surface. (Scaling may also occur in weak surfaces on concrete, resulting from poor finishing and curing practices.)

*(c) Leaching*
Porous or cracked concrete and mortar through which water is constantly flowing will lose portlandite (lime) and other soluble cement hydration products by dissolution. This will result in increased porosity and loss of strength. The dissolved ions are generally precipitated on the surface of the material in the form of a gel-like component. Leaching can occur both in concrete structures and in the mortar of brickwork structures. The rate of removal of hydration products will be increased by low pH (see Section AP I.2.2 (g)) or water containing aggressive carbon dioxide (see Section AP I.2.2 (h)).

*(d) I.2.3.4 Efflorescence*

Efflorescence is the deposition of salts (usually sulphates or chlorides) on the surface of a material. It can occur in both concrete and brickwork structures and is caused by the flow of water (as a liquid) through or from the interior of the material to the surface, where evaporation results in crystallisation of the dissolved salts. It is generally indicative of porous material and is often, though not always, associated with cracks.

The salts deposited on the surface and the exudations of lime described in Section AP I.2.3 (c) above, are frequently the most common indication (in some cases the only indication) of water ingress through the roof, walls or floor of a reservoir (see Figure AP I.2).

**Figure AP I.2** *Efflorescence on reservoir roofs and walls*

## AP I.3 CRACKING (OTHER THAN FORMED JOINTS)

Cracks are common in concrete. Many factors affect concrete that result in cracking and which influence the size of the cracks formed[28]. Cracks can also occur in the mortar and bricks of brickwork structures. The appearance and orientation of cracks can reveal a significant amount of information on their origin. Increase in the length and width of cracks (active cracks) or no significant change in crack status (passive cracks) is significant in assessing possible repair options.

> Many factors influence the effect of cracks on the integrity of the structure, and simple generalisations may be misleading. However, it is helpful to diagnose, as far as possible, the cause or causes of cracking which may be present in order that an effective repair can be achieved.

A number of different causes of cracking in reservoirs have been identified.

### AP I.3.1 Plastic shrinkage

Soon after casting, rapid evaporation at the surface of the concrete may exceed the rate at which water flows towards the surface from the core concrete, resulting in shrinkage of the surface concrete while it is still in a plastic state. The core concrete restrains this shrinkage, putting the surface concrete into tension to such a degree that cracks may form. The cracks are often wide but may be shallow and as such may be harmless unless the concrete surface comes into contact with aggressive species, e.g. chloride ions, when they provide a route for rapid ingress to the level of any reinforcement (see Section AP I.3.4). Once formed they may become the focus of tension strains due to other factors, and can widen to become a path for leakage.

### AP I.3.2 Drying shrinkage

These cracks result from the drying of the surface of the concrete after it has hardened (due to shrinkage of the cement hydrates). They are usually finer and deeper than plastic shrinkage cracks and may have a random orientation. They would normally be expected to occur on an unprotected face of an *in-situ* mass concrete roof exposed to a hot environment and/or drying winds. Alternatively, drying shrinkage cracks may occur as a result of restraint in the structure to shrinkage movement. They may provide a route for the rapid ingress of aggressive species.

### AP I.3.3 Plastic settlement

This would only be expected to occur at the top cast face of *in-situ* reinforced concrete that settles and bleeds excessively in the formwork. These cracks become visible in the first few days after casting, tend to form longitudinally, over the reinforcement and are more likely to occur in thicker slabs. They are a common cause of serious reinforcement corrosion (see Section AP I.3.4).

### AP I.3.4 Changes in stress state

Cracking resulting from changes in stress state can occur in all types of construction materials used in reservoirs and may be caused by increased loads, load relief, locally reduced section or material strength discontinuity, restrained thermal or moisture contraction/expansion or changes in material properties due to deterioration. Crack width varies, although orientation is usually well defined, e.g. diagonal cracks across the corners of reinforced concrete slabs (usually associated with torsional warping of the slab in conjunction with some level of fixity between the roof and the walls), arch thrust in brickwork roofs, mining or other forms of ground subsidence (detailed in Section AP I.3.10), settlement of poor/unsuitable fill (following modifications/additions to pipework below floors and/or walls), thermal or moisture induced movement and low strength concrete.

Biological factors, e.g roots from adjacent plants, can influence cracking by penetrating into existing cracks in the structure, increasing stresses and widening cracks (see Figure AP I.3).

During service, the width of individual cracks may vary depending on the water level within the reservoir. When the reservoir is drawn down for inspection, cracks may close only to open again during filling. If the cracking is sufficiently severe that structural integrity is considered to be in doubt, it would be advisable for a structural engineer to carry out an investigation.

### AP I.3.5 Sulphate attack

Random cracking due to the formation of expansive mineral phases within the concrete or mortar occurs. A more detailed description is given in Section AP I.2.2 (a)). The sulphate generally originates from the soil used for thermal insulation on roofs or from the groundwater.

### AP I.3.6 Reactive aggregate

Usually map cracking with a distinctive pattern caused by differential expansion of concrete (a detailed description of the process is given in Section AP I.2.2 (c)). Cracking increases in magnitude with severity of attack and time.

**Figure AP I.3** *Root penetration into a reservoir*

### AP I.3.7 Salt crystallisation
Random cracking and/or delamination of the material due to crystallisation pressure (a more detailed description of the process is given in Section AP I.2.2 (f)).

### AP I.3.8 Freeze-thaw
Closely spaced cracks or delaminations parallel to the surface of the concrete or mortar, resulting from the repeated freezing and thawing of cement paste critically saturated with water (see Section AP I.2.3 (b)). This mechanism has only been observed to occur where inadequate thermal insulation has been provided on a roof.

### AP I.3.9 Corrosion of reinforcement
Cracking is caused by the build-up of expansive corrosion products on the reinforcement. The cracking generally follows the pattern of the reinforcement. Migration of soluble corrosion products with crystallisation at some other location (generally where oxygen availability is higher) can occur (a more detailed description of the process is given in Section AP I.4).

### AP I.3.10 Ground settlement
Cracking due to ground settlement can be considered as being due to a change in stress state of an element (see Section AP I.3.4). Settlement may be widespread, affecting the entire reservoir, e.g. from mining subsidence, or be localised, e.g. from scour caused by outward leakage or compaction of fill above newly installed services into the reservoir. Settlement problems are usually manifested by joint movements. However, where this is not possible, cracking of the structure will occur. Orientation and location of the cracks will depend on the type, extent and location of the settlement. Diagonal cracks across the corners of floor slabs are typical.

### AP I.3.11 Thermal movement

Cracking resulting from thermal movement occurs in elements that are restrained at one end and which are exposed to large temperature variations: e.g. roofs where thermal insulation is insufficient or has been removed to undertake repair work, or walls when the reservoir has been drawn down for inspection. The extent of cracking will be dependent on the construction material and the degree and location of fixity.

## AP I.4 CORROSION OF REINFORCEMENT

The principal causes of reinforcement corrosion are:

- chemical attack
- chloride attack (which affects the stability of the oxide film even at high pH)
- loss of alkalinity of the concrete (reducing the pH of the concrete to a point where the passive film can no longer be maintained)
- porous cover concrete
- cracking (particularly plastic settlement cracks, which expose the reinforcement to the moist environment of the reservoir).

> The effects of these features are exacerbated by inadequate concrete cover. Lack of cover on columns, haunches and soffits was especially prevalent in early mesh reinforced structures.

Corrosion is the electrochemical oxidation of steel. It occurs when oxygen and moisture are present and when passivity of the metal has been lost. Passivity occurs as a result of the growth of a thin iron oxide film on the surface of the steel, stabilised by the high pH of the concrete environment. When steel corrodes, the products formed occupy a greater volume than the original metal, as much as 10 times greater in the case of the most highly oxidised types[29, 30]. This creates internal stresses, which will eventually overcome the tensile strength of the material resulting in cracking and spalling. Loss of section of the reinforcement in concrete may have significant structural consequences by reducing the ability of the structure to carry its design loads. Oxygen is required at the cathode to complete the corrosion cell. In fully saturated concrete (as would occur in the soffit of the roof slab, below water level walls and floor), the rate of oxygen diffusion may be low, and hence will control the rate of the corrosion process.

Low cover, poor quality concrete and plastic settlement voids below reinforcement bars are significant factors contributing to the onset of corrosion, reducing the protection offered by the cement paste and allowing aggressive ions to arrive at the surface of the reinforcement earlier. Any cracking of the concrete will also allow rapid ingress of aggressive ions and oxygen to the surface of the reinforcement.

## AP I.5 BREAKDOWN OF WATERPROOFING SYSTEMS

Two design philosophies have been adopted with regard to waterproofing reservoirs. The first relied on the materials from which the reservoir was constructed being watertight (impervious to water except under hydrostatic pressure) and therefore no additional measures were necessary, while the second allowed the construction of a non-watertight structure, which was then waterproofed (made impervious to water either in the liquid or vapour phase) by additional measures.

The most common methods of waterproofing involved the use of either a layer of puddled clay or sheet membranes, usually bituminous, bonded to the surface or unbonded and site welded to provide a continuous sheet.

Puddled clay waterproofing consists of a thin layer of natural clay, which when kept wet has a highly impermeable microstructure due to the grain size of the clay minerals. The clay is plastic but will fracture if sufficient movement occurs, or it can be penetrated by plant roots, which will allow free drainage of water from the surface. Puddled clay was more frequently used as waterproofing to reservoir walls, but when permitted to dry out, its waterproofing capacity is reduced and it shrinks, which could allow the walls (e.g. in masonry structures) to move outwards and possibly create problems in the roof.

Waterproof bituminous membranes (bonded or unbonded) may fail for a number of reasons. Hardening and shrinkage due to evaporation of volatile components and oxidation of the bitumen does not affect the waterproofing properties of the sheets; however, if they are then subjected to stress, cracking may result. This is less likely to occur in membranes reinforced with robust carrier films. The ability to survive induced stress decreases with increasing stiffness.

Bonded membranes may lose adhesion due to pressure build-up from moisture in the substrate, although they will still function effectively as unbonded membranes provided the joints between the sheets remain intact. The principal advantage of a bonded membrane is that the lateral movement of water across the substrate surface cannot occur in the event of damage to the membrane. The risk of loss of adhesion is increased where:

- Disintegrating type curing agents have been used.
- Laitance is present.
- Honeycombing or surface irregularities that leave cavities beneath the membrane occur.
- The primer is not allowed to dry sufficiently, or is permitted to accumulate in, or is not brushed out of surface depressions.

(See also Section B3.2.3.)

Unbonded membranes are susceptible to failure at the welded joint, usually due to poor site practice. This has serious consequences, as water may travel laterally beneath the membrane before ingressing the reservoir.

Membrane systems used (predominantly on reservoir roofs) based on bentonite work in a similar fashion to puddled clay. They require water to hydrate the hydrophilic material, which then forms a barrier to the passage of further moisture. Re-hydration, after drying, may only occur after lengthy periods, or may not occur at all, with a consequent loss in waterproofing ability, and could result in contamination of the water supply.

Liquid systems tend to be susceptible to weather at the time of application, workmanship and the quality of surface preparation, all of which may result in failure.

Many types of membrane system are susceptible to attack from organic solvents, notably petroleum products, and may be punctured by sharp objects if they are not protected.

Internal rendering has frequently been used for waterproofing. Failure can occur as a result of external or residual water pressure behind the render during drawdown, poor surface preparation, etc.

# Appendix AP II Ventilation requirements for reservoirs

## AP II.1 DERIVATION OF THE VENTILATION PROVISION FORMULA

This derivation is based on the 'worst case' scenario of gravity discharge of a full reservoir through a fractured outlet pipe at the base (see Figure AP II.1). The following terms apply:

$A$ = Surface area of the water in the tank.
$A_o$ = Outlet area.
$A_i$ = Inlet area.
$A_c$ = Area of the jet at the vena contracta.
$V_{res}$ = Volume of reservoir.
$V_a$ = Velocity of incoming air.
$V_{out}$ = Velocity of water at the outlet.
$w$ = Unit weight of water.
$h$ = Depth of water.
$H$ = Maximum water depth (Height of the reservoir).

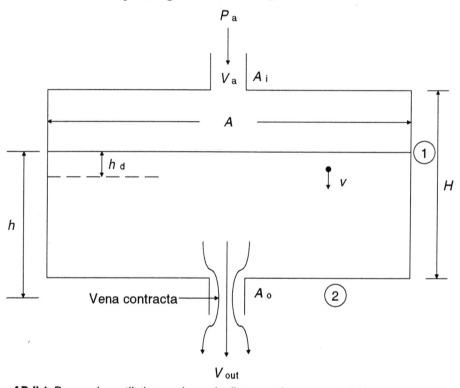

**Figure AP II.1** *Reservoir ventilation - schematic diagram of a reservoir drained by gravity*

Bernoulli's equation should be applied to the free surface and the vena contracta (1 and 2 respectively in Figure AP II.1):

$$(Z_1 - Z_2) + \frac{(P_1 - P_2)}{w} + \frac{(V_1^2 - V_2^2)}{2g} = 0$$

Where:

$(Z_1 - Z_2) = h$ = (the distance from the outlet to the vena contracta is very small compared with the depth of the reservoir).
$P_1 \approx P_2 = P_a$.
$V_1 <<< V_2$.

This allows a simplification of the above:

$$h - \frac{V_{out}^2}{2g} = 0$$

Where: $V_{out} = \sqrt{(2gh)}$.

In the case of a real fluid, there is a small energy loss due to viscous effects, which necessitates the introduction of a 'coefficient of velocity', $C_v$.

> Note 1: For most cases $C_v$ is typically 0.97.

The discharge:

$$Q = A_c V_{out}$$

It is more convenient, however, to express the discharge in terms of $A_o$, which is constant and easily measured. The areas $A_o$ and $A_c$ are related to the 'coefficient of contraction', $C_c$.

> Note 2: For most cases $C_c$ is typically 0.65.

Therefore:

$$Q = C_c C_v A_o \sqrt{2gh} = C_{do} A_o \sqrt{2gh} \tag{1}$$

Where: $C_{do}$ = the coefficient of discharge of the outlet of the reservoir.

> Note 3: For most cases $C_{do}$ is typically 0.63.

For compatibility, the volume of water discharged must equal the volume of air entering the tank assuming the air is incompressible over the small pressure changes being considered. The flow through the vents is given by:

$$Q = C_{di} A_i V_a \tag{2}$$

Where: $C_{di}$ = the coefficient of discharge of the inlets of the reservoir.

> Note 4: $C_{di}$ is dependent on the type and geometry of the particular vents being used and generalised values would be inappropriate.

Equating equations (1) and (2):

$$C_{do} A_o \sqrt{(2gh)} = C_{di} A_i V_a$$

For the worst case, i.e. initial drawdown ($h = H$) then:

$$A_i = \frac{C_{do} A_o \sqrt{(2gH)}}{C_{di} V_a} \tag{3}$$

The number of vents required is:

$$N = \frac{A_i}{A_{std}}$$

Where: $A_{std} \equiv$ the **effective** area of a standard vent.

Rearranging and substituting for $A_i$ in equation (3)

$$N = \frac{C_{do}A_o\sqrt{(2gH)}}{A_{std}C_{di}V_a} \qquad (4)$$

However, air velocity is dependent on the differential pressure across the vent and can be related using the laws of momentum:

$$Ft = mV_a$$

Where: $F$ = Force across the vent caused by the pressure differential.
$t$ = Unit time.
$m$ = Mass of air passing through the vent.
$V_a$ = Velocity of the air.

i.e.:

$$A\Delta P t = Q\rho_{air}tV_a$$

Hence:

$$V_a = \sqrt{\frac{\Delta P}{\rho_{air}}}$$

Where: $\rho_{air}$ = 1.23 kg/m³.
$\Delta P$ = Differential pressure drop across the vent.

Therefore:

$$N = \frac{C_{do}A_o\sqrt{(2gH)}}{A_{std}C_{di}\sqrt{\frac{\Delta P}{\rho_{air}}}} \qquad (5)$$

Let:

$$k_o = C_{do}A_o$$
$$k_i = C_{di}A_{std}$$

and rearranging for the general case:

$$N = \sqrt{\rho_{air}}\,\frac{k_o}{k_i}\sqrt{\frac{2gH}{\Delta P}}$$

$$N = \sqrt{1.23}\,\frac{k_o}{k_i}\sqrt{\frac{2gH}{\Delta P}}$$

$$N = 1.11\,\frac{k_o}{k_i}\sqrt{\frac{2gH}{\Delta P}}$$

For the specific case of a sudden discharge of water:

$$N = \sqrt{(2g\rho_{air})}\,\frac{k_o}{k_i}\sqrt{\frac{H}{\Delta P}}$$

$$N = \sqrt{(2\times 9.81 \times 1.23)}\,\frac{k_o}{k_i}\sqrt{\frac{H}{\Delta P}}$$

$$N = \frac{4.91\, k_o}{k_i} \sqrt{\frac{H}{\Delta P}}$$

This expression is related to the outlet and inlet discharge coefficients, the outlet pipe area, the standard vent area, water depth at commencement of drawdown and the pressure differential across the vents.

---

$\Delta P$ must be limited, by providing sufficient vents to prevent structural damage or more importantly to limit air velocity.

Pumping will change the rate of removal of water from the reservoir and may affect the pressure differential within it. However, the gravity model provides a worst case scenario representing fracture of the outlet pipe.

---

Example calculation: To determine the number of 150 mm diameter vents required in the roof of a service reservoir with a maximum water depth of 6.0 m above a 600 mm diameter outlet pipe. Assuming that the outlet pipe fractures and under these conditions the negative internal pressure must not exceed 1.0 kN/m², i.e. 1.0% of atmospheric.

For the purposes of this calculation assume $C_{do} = C_{di} = 0.63$ and $A_{std}$ is 50% of the mesh covered vent apertures:

$$N = \frac{4.91\, k_o}{k_i} \sqrt{\frac{H}{\Delta P}}$$

$k_o = 0.63 \times \pi \times 0.6^2/4 = 0.18$
$k_i = 0.63 \times 0.5 \times \pi \times 0.15^2/4 = 0.0056$

$$N = \frac{4.91 \times 0.18}{0.0056} \sqrt{\frac{6}{1000}}$$

$$= \underline{12}$$

CIRIA Report 138

# Part B

# Considerations in the decision making process

SECTION B1 SAFETY OF OPERATIONS
SECTION B2 DRAINAGE PROVISIONS
SECTION B3 AVOIDANCE OF DEFECTS ARISING FROM REPAIR ACTIVITY
SECTION B4 ROOF INSULATION AND EARTHWORKS
SECTION B5 DURABILITY, MAINTENANCE AND COST OF REPAIRS AND WATERPROOFING
SECTION B6 EXAMPLE MATERIAL PERFORMANCE REQUIREMENTS

# B1 Safety of operations

Everyday operations involving underground service reservoir repairs should ensure, so far as is practicable, the health, safety and welfare of the operatives and others who may be affected by the operations. In order to comply with these requirements it is necessary to specify and carry out a safe system of working.

## B1.1 PARTICULAR HAZARDS APPLICABLE TO WORKING IN AND AROUND SERVICE RESERVOIRS

Many of the hazards encountered during construction may also occur when working in and around existing service reservoirs.

Particular attention should, however, be paid to the following:

### B1.1.1 Hazards associated with working inside underground reservoirs
- confined space working, including

    - limited access (e.g. difficulties in rescuing injured personnel)
    - inadequate ventilation (e.g. oxygen deficiency/enrichment, toxic/noxious gases, explosions due to gases, fumes)

- working over water, including

    - flooding
    - drowning (e.g. due to undefined deep areas exposed on draw down).

### B1.1.2 Hazards associated with repair/waterproofing work
- handling of dangerous substances

- fumes and toxic/noxious gases.

## B1.2 RESPONSIBILITIES FOR SAFETY

There are two aspects to responsibilities for safety on construction sites.

- Health and Safety at Work Act, 1974 and other legislation.

    It is the employer's duty under the Health and Safety at Work Act, to provide such information, instruction, training and supervision as is necessary to ensure, so far as is reasonably practicable, the health and safety at work of his employees. In addition each employee is required to exercise reasonable care for the health and safety of himself and others who may be affected by his acts or omissions at work.

    Under the Health and Safety at Work Act 1974, consultants are obliged to do 'what is reasonably practicable' for safety. Consultants and clients are given greater responsibility under EC directives, which came into force on 31 March 1995, in the form of the Construction (Design and Management) Regulations (CDMR). These require the nomination of persons to safety roles, at both the design and construction stages. It is also necessary to formulate a health and safety plan, which must be initiated during the design stage of the scheme and the appointment of a Principal Contractor to ensure compliance with it.

- Conditions of contract.

Under most standard conditions of contract (refer to Section C3) the contractor is responsible for safety on site. For example, Clause 8 (3) of the ICE Conditions of Contract, 6th Edition states:

*'The Contractor shall take full responsibility for the adequacy, stability and safety of all site operations and methods of construction.'*

However, the designer/specifier has a definitive obligation to have considered all aspects of safety in his design: e.g. adequate site investigation, safe permanent works.

The contractor must be informed of all site investigation details and of known likely hazards during the tender stage, to permit him to plan safe site operations and methods of working.

Safe systems of work for specific hazards, such as working in confined spaces, are detailed in numerous documents, some in legislation and others as guidance documents. The contractor would be expected to be familiar with and implement them as part of his general obligations under the Health and Safety at Work Act, 1974.

## B1.3 PLANNING, PERSONNEL SELECTION AND TRAINING

### B1.3.1 Planning

In order to plan for safety, it is important to assess the scope and type of works to be carried out. This should involve:

- consideration of the works required
- the methods by which the work is to be carried out
- hazards, inherent in the plant, in the working methods proposed and in the working environment.

At this stage it is necessary to ensure that sufficient resources and time have been allocated to the task to enable a safe system of work to be followed.

### B1.3.2 Personnel selection

Personnel selected must be trained in, and experienced in, the type of operations involved and must be physically and mentally suitable. Similar care must be given to the selection of personnel nominated to act as attendants or rescue teams.

### B1.3.3 Training

Proper and effective training is necessary for:

- supervisors
- personnel delegated to enter confined spaces
- personnel delegated to act as attendants
- personnel delegated to form a rescue team.

Training must be appropriate to the situation and include:

- details of the hazards specific to the task
- the safe system of work and arrangements for its operation
- safe operation of equipment and a knowledge of its construction.
- instruction in the use of gas testing equipment, for work in confined spaces.
- how to deal safely with malfunctions and failures of equipment during use.
- works emergency procedures.
- instruction in first aid, treatment of shock, resuscitation, etc.
- instruction in the correct use and maintenance of rescue equipment.
- instruction and practice in the correct use of fire-fighting equipment.
- observance of personal hygiene rules to avoid health risks.
- use of communication equipment.

In addition to training, regular practices in rescue methods should be carried out by means of drills and simulations. Records of training should be retained and refresher training organised as necessary.

## B1.4 CONFINED SPACE WORKING AND PERMITS TO WORK

It is recommended that a permit-to-work system is followed for all operations that require entry into confined spaces. Adequate ventilation must be provided to maintain an atmosphere fit to breathe and to render harmless all dust, fumes or other substances that could be injurious to health.

If it is suspected that the atmosphere is not safe, it must be tested by a trained, competent, person. Personnel must not enter the confined space until the competent person is satisfied that the air is safe to breathe.

The permit to work is a written procedure, which ensures that the safe system of work is observed. It must be prepared by a responsible person who is familiar with the work procedures and has carried out a thorough assessment of the situation. It sets a safe procedure for entry, the correct sequence of work and the maximum time allowed in the reservoir. The typical sequence of operations for working in a confined space is shown in Figure B1.1[31]. The permit is issued by the person in control of the works when he is satisfied that conditions for safe working have been met.

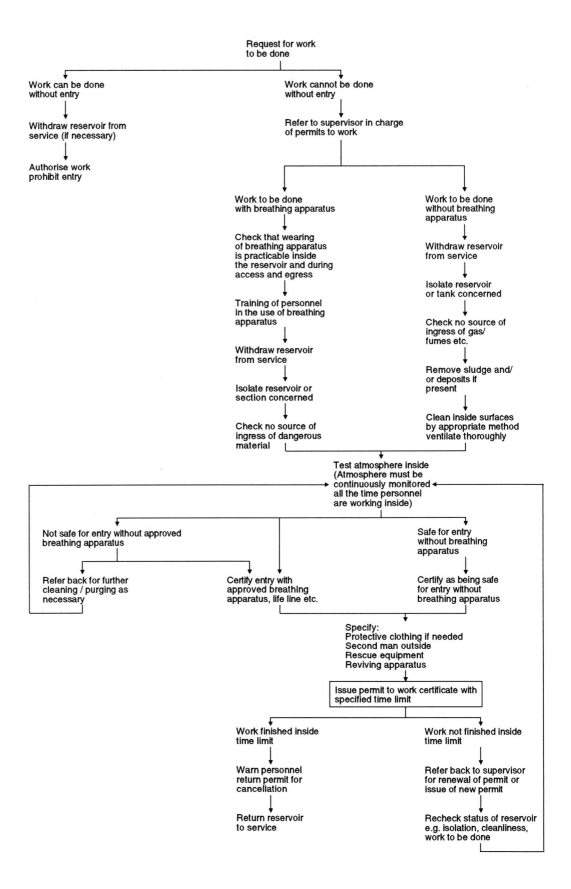

**Figure B1.1** *Typical sequence of operations for working in a confined space*

# B2 Drainage provisions

Adequate provisions for drainage should be made as part of the waterproofing work.

## B2.1 DRAINAGE PROVISIONS ON ROOFS

The necessity for waterproofing or repair of a reservoir roof usually arises from the need to prevent the leakage of contaminated water into the reservoir. The ponding of water on the roof increases the likelihood that leakage and contamination will occur. Problems of standing water on roofs generally arise from failure of the roof drainage system. This can result from a number of features including:

- blockage of existing drains
- insufficient drainage, especially in valleys of barrel and domed roofs
- deterioration of pipework
- inadequate falls on roof
- inadequate provision of details to convey water away from the edge of the roof and from access hatches and other openings.

Should water be permitted to pond on the reservoir roof then:

- The likelihood of leakage increases.
- There is a greater risk that standing water will become contaminated.
- Standing water may lead to unwanted plant growth.
- Dead load is increased, which may cause the structure distress.
- Hydrostatic pressure on waterproofing measures may increase the rate of penetration of water through any cracks or joints or through the intrinsic porosity of the material. In addition, hydrostatic pressure would be likely to increase the rate of leaching and thereby increase the width of any existing cracks.

---

Other features which need to be considered include the following:

- For reservoirs without sufficient fall across their roof, a fall could be provided by a screed laid to grade, (see Section A4.2.2), after checking the structural capacity. For vaulted roofs, localised screeds may be laid to form slopes in the valleys of the vaults.

- For reservoirs with inadequate fall or where it would be impractical to place a screed, then a drainage system should be provided over the roof area. Drainage systems[32], such as those listed below, should be designed to suit each particular reservoir roof:

    - perforated pipe (i.e. smooth plastic, corrugated plastic, porous concrete or open-jointed clay or concrete) surrounded by granular fill
    - single-size aggregate covered or surrounded by a geotextile filter
    - perforated pipe encased in a geotextile filter
    - preformed drains, i.e. filter fabric bonded to and surrounding an internal supporting mesh or dimpled sheet core
    - drainage tiles
    - polyethylene dimpled sheet cavity drain with a geotextile overlay.

Typical details of drainage provisions at the edge of a roof are given in Figure B2.1.

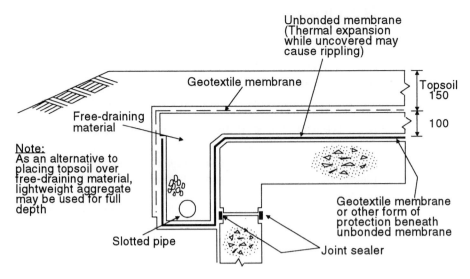

Detail A: For unbonded membrane

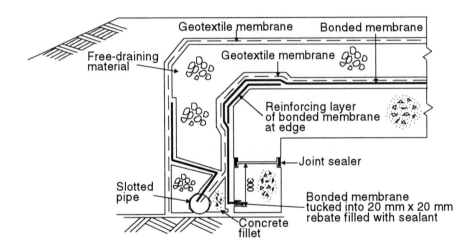

Detail B: For bonded membrane

**Figure B2.1** *Typical roof edge drainage details*

### B2.2 DRAINAGE PROVISIONS FOR WALLS

The two sources of water on the outside of reservoir walls are from roof runoff and from groundwater.

It is important that reservoirs are designed to prevent roof runoff water from flowing down the outside of the walls or from entering the reservoir through the roof/wall joint.

> It is recommended that collector drains are provided at the edge of the structure.

The consequences of continual water contact are similar to those listed for roofs and include the following.

- The likelihood of leakage into the reservoir increases.
- If the water is standing then the risk of contamination is increased.
- The hydraulic pressure can reduce the life of joint sealants and any other waterproofing measures.
- Hydraulic pressure will affect the stress state in the wall.
- The stability of the mound surrounding the reservoir could be undermined.

Particular attention should be paid to adequately conveying drainage water away from the reservoir structure and any surrounding mound.

Modern reservoirs are usually constructed above the water table, and if there is a risk of groundwater around the reservoir then a drainage layer is provided by single-sized aggregate, polyethylene dimpled sheet cavity drains or porous blockwork. Victorian reservoirs were sometimes built either adjacent to springs or with minimal regard for ground conditions. In such instances, leakage of contaminated water through walls can be a problem. The easiest method of dealing with such leakage is to seal the source of ingress from inside the reservoir (see Sections A3 and A4). However, where it is necessary to relieve the water pressure for structural or other reasons then the following options exist, although expensive (see Section B3.1.4).

- Reconstruct the earth mound incorporating a drainage layer and carrier drainage pipes.
- Excavate a trench around the wall perimeter, lay a carrier drain below floor level and fill the trench with single-sized aggregate.
- As above, but with the prior construction of a watertight sheet pile curtain. (As an alternative to the above, the excavated space between the sheet piling and the reservoir wall could be filled with concrete to act as waterproofing.)
- Should the reservoir be located on a perched water table then water could be carried to deeper draining strata by means of boreholes filled with drainage material.
- A watertight external wall could be constructed by the use of diaphragm walling techniques.

> Where extensive work is required to deal with water on the outside of the reservoir, the construction of a new reservoir should be considered, as this may be the cheaper option.

## B2.3 DRAINAGE PROVISIONS FOR FLOORS

The provision of underfloor drainage varies. It is often considered unnecessary unless uplift pressure could present a structural or stability problem. It could also be argued that a drainage system would prevent the build-up of stagnant water. The improvement or installation of underfloor drainage to an existing reservoir is not easy, and would almost certainly require the removal of large sections, if not all, of the floor. Suitable drainage systems would be similar to those listed in B2.1 for roofs.

> The provision of a lateral under-drainage system with regular inspection hatches will assist in monitoring leakage (by measuring dry weather flows when the reservoir is full) and identifying the source of leaks.
>
> Outward leakage through the reservoir floor can be detected by leaving a thin layer of water in the reservoir, then pressurising the drainage system with compressed air. The air will force its way back through the leaks, which can be located by the rising bubbles.

# B3 Avoidance of defects arising from repair activity

Prior to undertaking repair work, due consideration must be given to the possibility of additional defects, or problems arising as a result of the works and problems associated with materials used for the repairs. Examples of such occurrences are given in Sections B3.1 and B3.2.

## B3.1 STRUCTURAL PROBLEMS ARISING THROUGH REPAIR ACTIVITY

### B3.1.1 Cause: Drawdown of water – walls

*Example 1*
Problems may occur with walls when water or active earth pressure on the outside exceeds the hydrostatic pressure on the inside. Walls that are fixed at the bottom but not fixed to the roof may be forced to rotate about their base and move inwards at the top.

*Advice*: Check structural stability under the predicted (or measured) hydrostatic pressure.

*Example 2*
Bulging of brickwork linings due to the build-up of water pressure between mass concrete walls and brickwork linings (see also Example 4).

*Advice*: One solution is to provide temporary pressure-relieving taps at the bottom of the walls.

*Example 3*
When reservoirs that are waterproofed by puddled clay are left empty for any length of time, then the possibility exists that the puddle may dry out and shrink. When the reservoir is then refilled, the top of the wall would be pushed out beyond its original position and leakage may take place. The same may happen if embankments are permitted to dry out in hot weather. Should the roof be vaulted then spread may occur causing cracks at centre span (see Figure B3.1).

*Advice*: Check the structural stability prior to emptying the reservoir.

*Example 4*
Inward leakage of harmful substances through cracks may occur due to a hydrostatic pressure on the outside of the reservoir.

*Advice*: If the crack through which the leakage is occurring would be an in-service risk, it should be sealed.

### B3.1.2 Cause: Drawdown of water – floor

*Example 5*
In general, reservoirs are designed to withstand all the effects of drawing down and refilling, including the additional structural stresses induced when compartments of reservoirs are filled and emptied separately. However, it should be noted that when a reservoir is in service, the floor is evenly loaded. When it is emptied, the column footings will become more heavily stressed than the floor (reservoir floors may also be subjected to uplift due to groundwater pressure). This stress differential may be sufficient to either cause cracking in the floor slab (due to uplift) or in the puddled clay waterproofing layer under the floor. If this is the case then the reservoir may start to lose water upon completion of the works.

*Advice*: Check for structural stability prior to emptying the reservoir and aim to minimise the time period for which the reservoir is empty.

**Figure B3.1** *Cracks at the centre span of vaulted roofs*

### B3.1.3 Cause: Removal of roof insulation

*Example 6*
The removal of roof insulation will have two effects.

i)  The roof will become exposed to a different temperature and humidity regime.
ii) The roof will be subject to load relief.

These may result in a number of features:

1. Exposure to the weather

- General heating of the roof leading to overall expansion (see Figure B3.2). A similar effect would be seen when cold water is put into a previously warm empty reservoir. This may result in differential movement between the roof and walls if the roof/wall joint is debonded, or cracking if the joint is fixed.

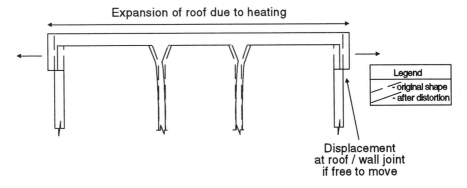

**Figure B3.2** *Expansion of roof due to heating*

- Heating or drying out of the top roof surface only. Both may lead to tensile stresses and thus cracking (see Figure B3.3) in the top surface.

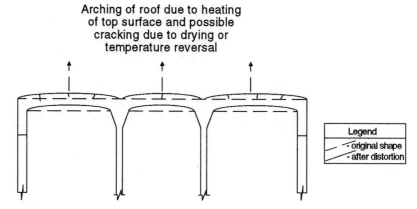

**Figure B3.3** *Arching of roof due to heating*

- Steel joists that butt against each other so as to form a single girder embedded in concrete may cause cracking due to differential expansion. A gap should therefore be left between them. Where the joists rest on the wall a gap should also be provided, which may be filled with asphalt so as not to crack the exterior wall.

- Damage to any waterproofing membrane on the roof.

2. Load relief

On removal of the insulation, load relief may lead to movement, the form of which will depend on the roof design and on the sequence of insulation removal. Some examples are as follows:

- A flat roof with no downstands may flex upwards and move outwards with respect to the wall. Cracking may appear on the top surface at centre span.

- An untied barrel arch roof would flex upwards. If the roof/wall joint is de-bonded then the roof may move inwards with respect to the wall (see Figure B3.4). If the roof/wall joint is fixed then cracking may ensue.

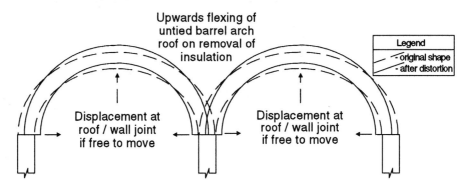

**Figure B3.4** *Upward flexing of untied barrel arches*

The sequence of insulation removal could also have a bearing. For instance, if removal is started at one edge and then moves across the roof to the opposite edge then those columns under the uncovered portion may restrain movement. This means that movement due to stress relief will be minimal near the last edge to be uncovered, but due to the cumulative effect will be a maximum at the starting edge.

*Advice*: Where possible, check for structural response prior to removing insulation, and work out a logical sequence for each structure so as to minimise movement and the risk of cracking.

### B3.1.4 Cause: Removal of Earth Embankments

*Example 7*
The removal of passive earth resistance from the walls may result in instability. In the past, reservoir walls were frequently designed so that the water pressure was counteracted by the earth backfill.

*Advice*: Where possible, check for structural response prior to removing the backfill, and establish a logical sequence of removal for each structure to minimise movement and the risk of cracking.

### B3.1.5 Cause: Construction activity

*Example 8*
The loading introduced by construction plant and stored materials may subject the roof to unacceptable stresses.

*Advice*: Acceptable loading from plant and materials should be calculated prior to commencement of construction activity. Preferably, limits on loading should be specified. Such restrictions may affect the method or order of working. Alternatively the roof should be propped by a temporary support system with stiffness adequate to pick up the live load.

*Example 9*
The spillage of undesirable fluids on the roof may lead to contamination entering the reservoir through unsealed cracks or joints in the roof or damage to an applied membrane.

*Advice*: No mixing of repair materials, solvents, adhesives, primers, etc., must be carried out on the roofs of reservoirs. No containers should be stored or left on the roof or roof drainage area, where they may leak or be knocked over.

*Example 10*
The spillage of diesel or petrol from plant used on the roof.

*Advice*: Consideration should be given to the banning or controlling of plant used on the roofs of reservoirs when they remain in service.

*Example 11*
Contamination of the reservoir – general (see also Section C1).

*Advice*: The following should be considered:

- All workmen to either complete a health questionnaire (with the understanding that if the Water Undertaker's doctor is not satisfied, the workman's own doctor can be consulted, or that further tests may be required) or to have been 'Widal tested' (a blood test for typhoid carried out by pathologists on specimens from all persons likely to contact potable water or potable water facilities).

- All personnel entering the reservoir should wash their boots in Chloros solution.

- The whole interior of the structure, particularly the repaired areas, should be pressure cleaned and disinfected before refilling. On completion of the disinfection no further entry should be permitted.

- The refilled, repaired reservoir will need a 24 hour bacteriological test before it can be returned to service.

- Depending on company policy, water in the refilled, repaired reservoir will need to be sampled and analysed for compliance with water quality parameters before being returned to supply.

## B3.2 PROBLEMS ASSOCIATED WITH REPAIR METHODS AND MATERIALS

The choice of the correct method and materials for repair or waterproofing of existing repair materials is as important as choosing a suitable system at the outset. The materials and techniques used must be appropriate in themselves, compatible with the type of material being repaired, and capable of overcoming any deficiencies in the original system. It is likely that the original cause of failure of the repair materials was due to their being subjected to forces or conditions more onerous than they were intended to withstand. Replacement of the affected area with similar materials or techniques may not prevent recurrence of the same problem.

Replacement of a structural element, as opposed to repair, is only normally justified when the stability of the element has been undermined, and it is so defective that satisfactory repair is likely to be either impossible, short-lived or uneconomic.

Determining the exact position of a defect can be difficult, as water may travel a long distance before appearing as a symptom in a position remote from the initial defect. For this reason, and also because there may be more than one type of defect, it is important that all possibilities are checked before reaching any conclusions. The method of repair should then be selected, having established the cause of the defect.

### B3.2.1 Cause: Problems with concrete repair systems

*Example 1*
Loss of adhesion between the concrete repair material and the substrate, due to shrinkage cracking or voids at the interface, etc.

*Advice*: Replace the concrete repair material with one compatible with the substrate, and with proper attention to preparation of the substrate: e.g. feather edging must be avoided. Attention should also be given to proper mixing, application and curing of the material. Random pull-off tests could be included as a part of the repair contract to check for adequate bonding.

### B3.2.2 Cause: Problems with screeds

*Example 2*
Insufficient fall of a roof towards drainage outlets.

*Advice*: Re-lay screed where necessary to correct fall.

### B3.2.3 Cause: Problems arising with waterproofing membranes on the outside

*Example 3*
Lack of adhesion to the substrate because of dusting or moisture between the membrane and substrate, or because of an uneven surface (see Figure B3.5).

*Advice*: Remove the membrane, and check that the moisture content of the substrate is within the tolerance of the material. Where the surface layers are friable or excessively dusty re-prepare the surface by grit-blasting and lay a new membrane. Where the surface is simply dusty, brush clean and apply a suitable surface primer prior to re-application of the bonded membrane. If the surface is simply too rough apply a levelling mortar. If the surface after preparation remains friable or degraded, apply an unbonded membrane.

**Figure B3.5** *A section of a membrane has been removed at a joint, showing adhesion failure between the membrane and substrate due to inadequate surface preparation, which led to leakage. Note also the selvage wrinkling present at laps.*

*Example 4*
Rippling, blistering, cracking, rucking or creasing in bonded membranes.

*Advice*: Check with the manufacturer if the material is defective. If this is not the case then the cause may be that the ambient temperature is too high and a more suitable grade of material should be selected. Alternatively, the laps between sheets can be made good by a sheet or liquid overlay and the membrane retained in place.

*Example 5*:
Rippling in bonded and unbonded membranes.

*Advice*: Remedy the cause, which may be vapour pressure beneath the membrane, moisture trapped between layers, material defects, thermal expansion of the membrane or movement of the structure. Many of these defects are aggravated by exposure to sunlight (see Figure B3.6), and therefore the membrane should be covered as soon as it has been pond or spray tested, or coated with a limewash, or similar, to provide a temporary reflective treatment if immediate loading is not possible. Defective areas in the membrane should be cut out and patched in an appropriate manner.

*Example 6*
Non-watertight overlap joints in a bonded membrane, because of a non-bonding edge strip, lack of adhesion or uneven material.

*Advice*: Replace the membrane with a more suitable material or make good the laps between sheets by the use of a sheet or liquid membrane.

**Figure B3.6** *Rippling in membranes*

*Example 7*
Backing sheet does not peel off the bonded membrane.

*Advice*: Wait until the temperature is within manufacturer's recommendations or replace the membrane with a more suitable grade of material. Check with the manufacturer if the material is defective.

*Example 8*
Melting of the adhesive under solar reflective backing sheet.

*Advice*: Cover the membrane as soon as possible with insulation and check for a defective batch. Stop work until weather conditions are within those recommended by the manufacturer. Check with the manufacturer if the material is defective.

*Example 9*
Mechanical damage to an existing membrane occurring during removal of insulation, or during laying of a new membrane.

*Advice*: Repair by cutting out the defective area and patch as recommended by the manufacturer.

*Example 10*
Surface crazing of membranes.

*Advice*: Treat by providing a solar reflective covering.

### B3.2.4 Cause: Problems with internal waterproofing treatments

*Example 11*
Lack of adhesion to the substrate.

*Advice*: Check that the moisture content of the substrate is within the tolerance of the material. Treat the surface of the concrete to remove dust and friable material.

*Example 12*
Mechanical damage.

*Advice*: Repair by preparing the damaged area and surround, and reapply a compatible material.

### B3.2.5 Cause: Problems with surface coatings

*Example 13*
Lack of adhesion to the substrate because of moisture, dusting or an irregular surface.

*Advice*: Check with manufacturer if the material is defective. Treat the substrate surface to remove moisture, dust and all friable material.

### B3.2.6 Cause: Problems with metalwork paint systems

*Example 14*
Lack of adhesion.

*Advice*: Check that the substrate moisture level is within the tolerance of the material and that surface preparation has been undertaken properly. Check with the manufacturer if material is defective. If the surface moisture content cannot be reduced specify a material with a higher moisture tolerance.

### B3.2.7 Cause: Problems with crack sealants

*Example 15*
Lack of adhesion to the substrate because of dusting or moisture in the concrete.

*Advice*: Remove the existing sealant and prepare the substrate properly including priming. If moisture is the problem then make sure that the concrete is dry before the next application or select a sealant which is moisture tolerant.

*Example 16*
Cracks showing through the material used for crack sealing (see Figure B3.7).

*Advice*: The crack is still moving and should be treated as such!

### B3.2.8 Cause: Problems with joint fillers and sealants

*Example 17*
Lack of adhesion to the substrate because of dusting or moisture in the concrete.

*Advice*: Remove the existing sealant and prepare the substrate properly. If moisture is the problem then ensure that the concrete is dry before the next application, or select a sealant and primer that are moisture tolerant.

**Figure B3.7** *Recracking of a 'sealed' crack*

### B3.2.9 Cause: Problems with Waterstops

*Example 18*
Displacement of a waterstop during a pour.

*Advice*: It is easier to prevent displacement of the waterstop occurring than to rectify a defective waterstop at a later date. Care should be taken during initial installation to ensure that the waterstop is correctly fixed, that placing the concrete does not displace the waterstop, and that compaction of the concrete around the waterstop is adequate. Where additional security is required, tubes may be incorporated into the joint for later remedial grouting should leakage occur.

# B4 Roof insulation and earthworks

## B4.1 ROOF INSULATION

An important part in the design of the **method of repair** to be adopted for a roof is the choice, for any given reservoir, of insulation. It is not possible to recommend any particular type of insulation, as the most suitable will depend on a number of factors, including the type of waterproofing system adopted (see, in particular, Section A4.2.3 (c)). However, the following points set out some of the advantages and disadvantages of alternative insulation materials.

> If the type of insulation on a reservoir is to be changed, with a consequent change in the appearance, planning permission is likely to be required if the roof is visible.

### B4.1.1 Seeded earth

When earth is used for insulation it is usually placed over a geofabric or a layer of pea gravel or similar single-size aggregate. This then provides a drainage layer under the soil.

*Advantages*
Environmentally pleasing if the reservoir is in public view, although many reservoirs are fenced for security reasons and are not intended as public areas. Maintenance (grass cutting) is simple. The site, if allowed to develop by restricting grass cutting, will attract wildlife and wild plants, native to the surrounding area. Many reservoirs have become havens for rare plants, birds, butterflies, etc., improving the Water Undertaker's image with regard to conservation.

*Disadvantages*
- Requires grass cutting.
- Requires a minimum soil depth of 150 mm, preferably 200 mm, to enable grass to grow.
- Will inevitably attract dog walkers, playing children and grazing animals unless securely fenced off.
- Membrane will need to be deep enough or sufficiently protected to avoid puncture from cricket wickets, etc.
- When grass is to be removed and stored for subsequent replacement after a repair contract, it should initially be killed and the turf stacked in piles of less than 1.5 m in height.

### B4.1.2 Gravel

It is recommended that either 20 mm single-size washed carboniferous limestone, 40 mm single-size washed aggregate, or lightweight aggregate are used for insulation, as none of these encourage the growth of plants. A minimum depth of 100 mm should be used although it may be necessary to provide 150 mm or more if insulation properties are required.

*Advantages*
- Environmentally pleasing if well designed and maintained.
- Will not require grass cutting.
- Rounded aggregate minimises damage to waterproofing membranes.

*Disadvantages*
- May require weeding (with certain sizes of aggregate).
- May attract motor and pedal cyclists.
- Not generally suitable in urban locations, unless adequately fenced off.
- Aggregate can provide ammunition for vandals.
- Sea-dredged aggregate will require washing to remove chlorides before use.

### B4.1.3 Other

As an alternative to the above, the reservoir roof may be used for a specific activity, e.g. sports facilities, car parking. However, should a policy of permitting access for the general public be adopted, then appropriate security measures will need to be taken to ensure that the reservoir is contamination-proof for all circumstances. The structural capacity of the roof should also be checked to ensure that it is adequate.

## B4.2 EARTHWORKS

The most common wall surround to covered reservoirs is an earth embankment, unless the excavation was on the side of a hill.

There are a number of factors to consider.

- In some cases a puddled clay layer was laid against the walls to form a waterproofing layer; otherwise the earth slope plays no part in waterproofing.

- Should the reservoir be losing water through the walls then this could weaken the earth surround, leading to failure of the embankment.

- Considerable excavation of the earth surround could be necessary either for the repair of walls (on the outside) or for the installation of peripheral drainage. When such excavation is planned, the reservoir structure must be checked for stability under this temporary stress condition, as in some previous designs passive earth resistance was used to counteract water pressure or the thrust from arched roofs.

- Any excavation must be careful not to damage drainage layers, pipes, granular fill, porous blocks or proprietary drainage sheets.

- Many service reservoirs have a strip of imported granular fill between the walls of the structure and the earth embankment. Its function is to fill the gap should earth movement take place, thus ensuring uniformity of loading against the structure. Any work on the embankment should maintain the integrity of the strip.

# B5 Durability, maintenance and cost of repairs and waterproofing

The selection of a **method of repair** for a specific reservoir defect will be a function of:

- technical assessment
- durability
- maintenance
- cost.

The relationship, and order of importance, between the above properties is a matter for consideration by the individual Water Undertaker. It is the objective of this section to introduce the concepts of durability, maintenance and cost. The technical assessment of the **methods of repair** is described in Section A3.

## B5.1 DURABILITY OF REPAIRS AND WATERPROOFING

The limit of durability is reached when an item fails to perform, or requires unacceptable expenditure on maintenance, to sustain its performance. In order to predict the durability of an item, it is necessary to identify the factors that are likely to have an effect, to measure the extent of their effect and the time for deterioration to reach the durability limit.

Factors that may affect the service life of materials used in reservoir construction and repair are: temperature, solar radiation, water, oxygen, carbon dioxide, acid gases, inorganic contaminants, biological contaminants (e.g. from microorganisms, insects, other animals, plants), stress, chemical and physical incompatibility and use in service. Investigation of the effect of these factors is described in Section A4.

The problems associated with assessing the durability of repair and waterproofing materials are as follows.

1. Most proprietary repair materials have not been in existence for more than 10–20 years and long-term experience of *in-situ* testing is therefore generally not available.

2. Older materials may not be representative of their modern counterparts as formulations change.

3. Accelerated testing of materials may not necessarily represent their actual performance in use.

The following general observations can be made.

- It is often possible to obtain a product guarantee for a proprietary material from the manufacturer usually for periods of up to 10 years. For the guarantee to be valid, the supervising engineer must ensure that the material was applied strictly in accordance with the manufacturer's recommendations and is within the scope of the recommended usage. The installer will provide the guarantee for the finished work. Insurance-backed guarantees are available for extended time periods from some organisations.
- Estimates of the ranges of expected service lives for a number of materials are given in Table B1. All the lives quoted assume that the material has been correctly applied and was fit for purpose after application.

> Additional information on the concept of durability of buildings and components is given in BS7543[33], which could usefully be applied to reservoirs.

**Table B5.1** Indicative estimates of the durability of materials

| Material category | Expected service life (years) |
| --- | --- |
| Concrete repair systems | 20 - 100 |
| Waterproofing membranes | More than 20 |
| Surface coatings | 1 - 15 |
| Metalwork paint systems | 5 - 10 |
| Crack injection | 20 - 100 |
| Crack / joint sealant | 10 - 20 |

## B5.2 MAINTENANCE

The purpose of maintenance is to extend the durability limit of a material or part of a structure. Maintenance of reservoirs falls into three stages: inspection, monitoring and repair work. Inspection is the first stage and should be undertaken as regularly as is practicable and with as much detail as possible being recorded during each visit. Inspection methods are described in Section C2.

After an inspection, the results should be compared with those from previous inspections and any changes in the state of the reservoir, or development of defects, noted. If these changes are considered to be sufficient to cause concern for structural integrity, water quality, etc., then regular monitoring, or repair when appropriate, should be considered.

Monitoring can take the form of regular visual inspections, increased frequency of water sampling and analysis, physical measurements if movement is suspected or testing of materials where deterioration, or a change in the state, of the structure is suspected. Monitoring techniques would essentially be the same as the methods for investigation detailed in Section C2.

Repair work would then take place when required as a result of the inspection and monitoring process. Initially, the repair of small areas may be all that is required. However, over a period of time it will probably become more cost-effective to renew the item or undertake more substantial repairs. For example, consider a reservoir roof that is constructed of precast and *in-situ* sections, where the joint sealants are beginning to deteriorate. A typical maintenance programme may involve the localised removal of insulation and replacement of the sealant, where failure has occurred, on a regular basis. Such piecemeal work would inevitably become quite costly and it may be more economic, in the long term, to cover the roof with a waterproofing membrane in a single contract.

Details of methods to be used for the maintenance of proprietary repair materials should be requested from the manufacturer at the time of installation, to prevent unnecessary failures.

## B5.3 COST

A large number of reservoirs in this country are more than 100 years old and still fit for purpose. It is therefore likely that many structures will outlive the service life of the materials used for waterproofing and repair work. Careful consideration should therefore be given to the estimated future cost, associated with the durability, of the chosen solution and the likely maintenance programme.

In many cases there will be a number of different options available for both the **method of repair** and the materials and products available. The durability of each option should be assessed and the long-term cost of possible maintenance and renewal considered during initial selection.

The following is a check list for some of the items (other than general preliminaries and overheads) which may make up the cost of repair work.

- operational cost of taking a reservoir out of service for a period of time, including cleaning and sterilisation
- provisions for safe working (see Section B1)
- investigation and testing
- lighting and other power requirements
- access ladders and scaffolding
- removal, stockpiling and replacement of topsoil and insulation from the roof
- excavation for peripheral drainage
- removal of existing access hatches and their replacement with new
- refurbishment of ventilation openings
- improvements to roof cross-falls and drainage provisions
- preparatory work for the purchase and application of proprietary repair/waterproofing materials.

Typical estimated unit costs for applying waterproofing and repair materials are given in Table B2. The costs include surface preparation, saw cutting, concrete removal, primers, materials, lightweight plant and labour necessary for product application. It is assumed that areas to be repaired/waterproofed are readily accessible.

The costs do not include contract preliminaries or any of the items listed above except the last item.

**Table B5.2** Typical estimated applied unit costs of waterproofing and repair materials

| Material category | Subcategory | Generic group | Estimated cost (see also Notes 1 & 2) |
|---|---|---|---|
| Concrete repair systems | Hand placed | Polymer modified cementitious (50 - 75 mm deep repair) | £150 - £300 /m² |
| | Sprayed concrete | Polymer modified cementitious (50 - 75 mm deep repair) | £100 - £250 /m² |
| | Flowable concrete | Polymer modified cementitious (50 - 75 mm deep repair) | £150 - £250 /m² [3] |
| Screeds | | Styrene butadiene mortar (5 mm) | £15 - £25 /m² |
| | | Epoxy mortar (5 mm) | £35 - £55 /m² |
| Waterproofing membranes (outside) | Bonded | Hot applied bitumenized fabric (3.5 mm) | } |
| | | Cold applied bitumenized fabric (2 mm) | } £15 - £20 /m² |
| | | Polyethylene bitumenized fabric (1.5 mm) | } |
| | Unbonded | Low density polyethylene (1000 μm including protective felt) | £5 - £10 /m² |
| | | Sodium bentonite sandwich (2 mm) | £15 - £20 /m² |
| | Liquid | Polyurethane | £18 - £35 /m² |
| | | Flexible cement slurry (3 mm) | £15 - £25 /m² |
| Internal waterproofing treatments | Sprayed concrete | Polymer modified cementitious | £30 - £40 /m² |
| | Waterproof slurry | Polymer modified cementitious | £20 - £30 /m² |
| Surface protective coatings | | Epoxy resin | £15 - £30 /m² |
| Metalwork paint systems | | Bituminous<br>Chlorinated Rubber<br>Epoxide<br>Polyurethanes | £20 - £30 /m² |
| Crack injection | | Epoxide<br>Latex emulsion<br>Polyurethane<br>Polymer modified cementitious | £40 - £140 /linear m |
| Crack sealants | Flexible sealant | Rubber bitumen | £10 - £15 /linear m |
| | | Polysulphide | £15 - £20 /linear m |
| | Preformed strip sealant | Polychlorinated rubber (200 mm wide) | £30 - £40 /linear m |
| Joint fillers and sealants | Joint fillers | Non - absorbent polyethylene | £35 - £40 /m² |
| | Joint sealants | As for crack sealants | £10 - £20 /linear m |
| | preformed strips | Polychlorinated rubber (200 mm wide) | £30 - £40 /linear m |
| Large volume grouting | | Cementitious<br>Resin | £150 - £300 /m³ |

Notes: (1) All costs given represent January 1994 prices.
(2) See Section B5.3 for the items included in the make-up of the quoted costs.
(3) Cost does not include shuttering.

# B6 Example material performance requirements

Currently there are no national or international specifications, codes or standards covering the performance requirements of repair materials and techniques. The example performance requirements presented in this section are drawn from experience and **individual clauses should be selected and developed by Water Undertakers to ensure suitability for the particular reservoir application** before being used for contractual purposes. Contractual specification of materials should be written:

> '...with a clear and realistic concept of the way they can be enforced. This implies that it should be possible to check whether every requirement is being met and if any breach of the requirements is discovered, suitable remedial action can be demanded. The action to be taken should be related to the probable consequences of the lack of compliance with the requirements...'

Taken from BSI Publication, PD6472: 1974 – *Specifying the Quality of Building Mortars.*

Materials covered by this section are:

B6.1 Concrete repair systems
B6.2 Waterproofing membranes on the outside
B6.3 Internal waterproofing treatments
B6.4 Surface protective coatings
B6.5 Metalwork paint systems
B6.6 Crack injection
B6.7 Crack sealants
B6.8 Joint sealants
B6.9 Large volume grouting.

For several of the performance requirements in the following sections there may be more than one appropriate test method. All testing should be undertaken by an approved laboratory which should, where possible, operate an accepted quality assurance scheme, e.g. BS5750, or be NAMAS accredited. Results from such tests may be acceptable but guidance should be sought from the manufacturer of the material where specific test methods are not given, a suitable test method being approved by the Water Undertaker.

In some instances, minimum quantitative requirements are quoted. The selection of values for inclusion in a specification for a specific project is subjective and depends on the particular application. The values quoted are generally taken from product literature as representing the minimum order of performance that can be expected from materials currently in use.

The legal requirements and approval schemes relating to materials in contact with potable water are outlined in Sections A4 and C1 respectively. It is the responsibility of the Water Undertaker to determine the acceptability of any particular type of material prior to its being specified and used.

## B6.1 CONCRETE REPAIR SYSTEM

> For structural repair, where load sharing by the repair material is anticipated, the repair material properties should be matched as closely as possible to those of the substrate concrete (as determined during the investigation; see Section C2). Modulus of elasticity, coefficient of thermal expansion and creep are important factors requiring consideration to prevent property mismatch.
>
> For polymer modified cementitious repair materials the test methods quoted (from British Standards BS1881 for concrete, or those from BS6319 for resin and polymer/cement compositions, but not as yet amended to completely cover these materials) may require some modification, e.g. curing requirements, sample size, etc. Where any doubt exists guidance should be sought from the manufacturer.

Examples of the minimum performance requirements for repair materials are as follows.

1. Adhesion to substrate

    The adhesive strength of the repair material to the concrete, and the cohesive strength of the repair material shall be a minimum of 1.5 N/mm$^2$. Tests can be undertaken either on repaired areas on site, in accordance with BS1881: Part 207: 1992, or on simulated repairs in the laboratory, in accordance with BS1881: Part 207: 1992 or BS6319: Part 4.

2. Shrinkage

    The total shrinkage of the repair material at any age up to a maximum of 6 months, tested at 20 °C and at 50% relative humidity shall not be greater than 0.04% when tested in accordance with BS1881: Part 5: 1970.

> Many proprietary repair materials contain shrinkage-compensating admixtures as a part of their formulation.

3. Compressive strength (where structurally appropriate)

    The compressive strength of the repair material shall not be less than that of the parent concrete when tested in accordance with BS1881: Part 116: 1983 or BS6319: Part 2: 1983.

4. Tensile strength

    The tensile strength of the repair material at 28 days shall be equivalent to or greater than that of the parent concrete when measured in accordance with BS1881: Part 117: 1983 or BS6319: Part 7: 1985.

5. Modulus of elasticity

    The 28 day elastic modulus of the repair material shall lie within the range of 10 to 25 kN/mm$^2$, for non-structural repairs, or shall be matched to that of the substrate concrete (probably in the range 25 to 30 kN/mm$^2$) for structural repairs when tested in accordance with BS1881: Part 121: 1983 or BS6319: Part 6: 1984.

6. Coefficient of thermal expansion

    The coefficient of thermal expansion shall match that of the substrate concrete or shall lie within the range of 8 to 15 × 10$^{-6}$ /°C.

7. Permeability

   The coefficient of permeability of the repair material to chloride ions shall be less than $1 \times 10^{-12}$ m/s when subjected to a pressure equivalent to 1 m head of water when tested in an approved permeability rig.

8. Chloride ion diffusion

   The coefficient of chloride ion diffusion resistance shall be less than $1 \times 10^{-12}$ m²/s.

9. Water immersion

   The repair material shall be suitable for permanent immersion in water.

10. Potable water

    All components of the repair system which could potentially come into contact with potable water shall comply with Regulation 25 of the Water Supply (Water Quality) Regulations 1989, as amended.

> This may require the Water Undertaker to make an assessment of the risk implied by the use of a particular material and under which section of Regulation 25 (1) to obtain, or give, approval (see also Sections A4.1 and C1.3.1). In practice, for new materials:
>
> for **large area application** - it is likely that all components of the repair system should be approved by the Secretary of State or be included in the list of approved substances, products and processes issued by the Secretary of State. (Approval under Regulation 25 (1) (a) or (d).)
>
> **or**
>
> for **small area application** - it may be acceptable for all components of the repair system to be Water Byelaws Scheme approved products, listed in the WRc *'Water Fittings and Materials Directory'* as evidence of having satisfied the requirements of BS6920: Part 1: 1988. (Approval under Regulation 25 (1) (b).)
>
> For materials used prior to 6 July 1989, approval could be given under Regulation 25 (1) (c).

## B6.2 WATERPROOFING MEMBRANE ON THE OUTSIDE

Examples of the minimum performance requirements for all types of waterproofing membrane (i.e. bonded, unbonded, hydrophilic and liquid) are as follows (the test methods specified in (1) to (6) are not appropriate to hydrophilic membranes and advice from the manufacturer should be obtained).

1. Watertightness

   The membrane shall be completely waterproof under a pressure equivalent to a 1 m hydrostatic head when tested in accordance with BS3424: Part 26: 1986 for roof application, and to a hydrostatic head equivalent to the full height of the walls plus 1 m when tested in accordance with BS3424: Part 26: 1986 for wall and floor application.

2. Chemical resistance

   The material shall be resistant to concentrations of chemicals considered appropriate to the specific project, during and possibly after construction, but which may include: dilute acids, dilute alkalis, soft water, sulphates and chlorides.

3. Water immersion

   The membrane shall be suitable for permanent immersion in water. Swelling of the membrane shall be less than 1% when tested in accordance with BS2782: Part 8: Method 830: 1986.

4. Application temperature

   As considered appropriate for the specific project.

5. Resistance to ultraviolet light

   The membrane, or the membrane plus any recommended protection, shall be resistant to 24 hours exposure to ultraviolet light (hours of exposure should be adjusted to be appropriate for the specific project). The membrane shall not lose flexibility in this time when tested in accordance with BS2782: Part 5: Method 550A: 1981.

6. Potable water

   The membrane system (including primers, etc.) shall comply with regulation 25 of the Water Supply (Water Quality) Regulations 1989, as amended (see also the note attached to Section B6.1 (10)).

### B6.2.1 Bonded preformed membranes

In addition to requirements (1) to (6) above, examples of the additional performance requirements for a bonded membrane are as follows.

7. Ability to span cracks

   The membrane shall be capable of spanning cracks of up to 3.0 mm in the substrate without tearing or being damaged in any way. In addition, the membrane shall have the capacity to span unanticipated cracks of 0.6 mm that occur after laying of the membrane.

8. Puncture resistance

   Once placed, the membrane shall be resistant to puncturing, creasing, de-bondment and any other damage from rubber-soled boots and light, rubber-tyred, vehicles such as wheelbarrows. The material shall have a minimum puncture resistance of 220 g when tested to BS2782: Part 3: Method 352D: 1979.

> The actual puncture resistance specified is subjective and advice should be obtained from manufacturers of suitable materials.

9. Adhesion

   The substrate shall be prepared and membrane bonded in accordance with the manufacturer's instructions. The membrane shall achieve a continuous bond to the substrate surface, such that the lateral movement of water beneath the membrane is prevented. The membrane shall, where possible, be covered with insulating material immediately after laying to prevent blistering. Where covering is not feasible a temporary reflective coating (e.g. lime wash) should be applied to the membrane surface.

### B6.2.2 Unbonded preformed membranes

In addition to requirements (1) to (6) above, examples of the additional performance requirements for flexible sheet-type unbonded membranes are as follows.

(1) to (6) plus:

7. Puncture resistance

    The membrane shall be resistant to puncturing and damage from rubber-soled boots and light rubber-tyred, vehicles such as wheelbarrows. Puncture resistance shall be greater than 290 g when tested to BS2782: Part 3: Method 352D: 1979.

> The actual puncture resistance specified is subjective and advice should be obtained from manufacturers of suitable materials. The suggested value is greater for unbonded systems than for bonded systems owing to the greater risk of water travelling beneath the membrane before penetrating into the reservoir (i.e. the point of leakage into the reservoir may not be the point of leakage through the membrane, hence safety factors are increased) and the consequential increase in costs of investigation to locate the damage to the membrane.

8. Ultimate strength at fibre breakdown

    Ultimate strength at fibre breakdown shall be greater than 11 kN/m when tested in accordance with BS2782: Part 3: Method 327A: 1982 (1988).

9. Elongation

    Elongation at break shall be greater than 100% at 23 °C when tested in accordance with BS2782: Part 3: Method 327A: 1982.

### B6.2.3 Unbonded hydrophilic membranes

> Active, sodium bentonite membranes, once wetted, must thereafter be prevented from drying out. Failures have occurred as a result of drying, particularly during periods of low rainfall. Experience has shown that the application of a selected granular material placed directly over the membrane will help to retain water and prevent drying out.

In addition to requirements (1) to (6) above (excepting the swelling requirement (3)), examples of the additional performance requirements for hydrophilic membranes are as follows.

(1) to (6) plus:

7. Composition

    The active hydrophilic material shall be sodium montmorillonite hydrous aluminium silicate: i.e. natural sodium bentonite, having a permeability of less than $1 \times 10^{-9}$ litre/m$^2$/s.

8. Surface preparation

    The substrate shall be free from vegetation and debris with no surface protrusion exceeding 10 mm.

    Membranes may be laid horizontally on a prepared sand substrate, which shall consist of a minimum of 25 mm of clean washed sand complying with BS882, compacted to exhibit at least 85% of the maximum dry density derived by the Proctor test, with a pH in the range 5 to 8 and finished to falls.

    Brickwork or blockwork walls may be rendered with a sand:cement mix prior to placing the membrane.

9. Installation

    The prepared surface shall be free from standing water and/or ice. The membrane shall be placed in accordance with the manufacturer's recommendations. Immediately after each strip has been laid the insulation shall be installed.

10. Ability to span cracks

    The membrane shall be able to span existing cracks up to 0.3 mm in the substrate without tearing or being damaged in any way. In addition the membrane shall have the ability to self-heal cracks of 0.6 mm or increases of 0.6 mm in existing cracks that occur after laying of the membrane.

11. Joints

    All joints shall be perpendicular to the slope with 300 mm minimum laps.

12. Reinstatement of insulation

    The minimum depth of insulation shall be 300 mm and shall consist of:

    (a) 100 mm layer of 12 mm single-sized washed rounded gravel (maximum particle size 18 mm) complying with BS882 in direct contact with the membrane

    (b) polypropylene geotextile non-woven fabric with a permeability greater than 50 litre/m$^2$/s at a 100 mm hydrostatic head

    (c) gravel or topsoil a minimum of 200 mm in thickness.

    No machinery or equipment shall move over the bentonite membrane without the 300 mm depth of insulation in place.

13. Drainage

    Land drains shall be installed at a maximum of 35 m centres to discharge into a perimeter drainage system for installations with edge dimensions exceeding 35 m in length.

### B6.2.4 Liquid membranes

In addition to requirements (1) to (6) above, examples of the additional performance requirements for liquid membranes are as follows.

(1) to (6) plus:

7. Ability to span cracks

    The membrane shall have the capacity to span cracks of up to 2.0 mm in the substrate. In addition, the membrane shall have the capacity to span unanticipated cracks of 0.6 mm that occur after the membrane has been applied.

8. Adhesion

    Bond strength to the substrate, which has been prepared in accordance with the manufacturer's instructions, shall be greater than 2.0 N/mm$^2$ when tested in accordance with ASTM D 4541–85. The membrane shall remain bonded to the substrate under the pressure of water vapour coming through the concrete.

9. Impact resistance

The membrane shall have an impact resistance of greater than 15 J when tested in accordance with BS3900: Part E7: 1974 (1986). The membrane shall be tested 7 days after application, or after the manufacturer's recommended curing period, whichever is greater.

## B6.3 INTERNAL WATERPROOFING TREATMENTS

Examples of the minimum performance requirements for all waterproofing treatments (liquid membranes, coatings or renders) on the inside surfaces of reservoirs are as follows.

1. Watertightness

    The material shall be completely waterproof under pressure equivalent to a 1 m hydrostatic head when tested in accordance with BS3424: Part 26: 1986 for roof application, and to a hydrostatic head equivalent to the full height of the walls plus 1 m when tested in accordance with BS3424: Part 26: 1986 for wall and floor application.

2. Chemical resistance

    The material shall be resistant to concentrations of chemicals considered appropriate to the specific project, during and possibly after construction, but which may include: dilute acids, dilute alkalis, soft water, sulphates and chlorides.

3. Water immersion

    The material shall be suitable for permanent immersion in water.

4. Application temperature

    As considered appropriate for the specific project.

5. Potable water

    The material (including primers, etc.) shall comply with Regulation 25 of the Water Supply (Water Quality) Regulations 1989, as amended (see also the note attached to Section B6.1 (10)).

6. Adhesion

    The bond strength to a substrate prepared in accordance with the manufacturer's recommendations shall be not less than 1.5 N/mm$^2$. The material shall have the capacity to remain bonded to the substrate under a hydrostatic head of 1 m on the substrate side of the membrane for roofs, and to a hydrostatic head equivalent to the full height of the wall plus 1 m for walls and floors.

### B6.3.1 Liquid membranes

In addition to requirements (1) to (6) above, example performance requirements for liquid membranes are as follows.

(1) to (6) plus:

7. Ability to span cracks

    Membranes shall be able to span cracks of up to 2.0 mm in the substrate. In addition, the membrane shall have the capacity to span unanticipated cracks of 0.6 mm that occur after laying of the membrane.

### B6.3.2 Polymer-modified cementitious coatings, mortars and renders

In addition to requirements (1) to (6) above, example performance requirements for polymer-modified cementitious coatings, mortars and renders are as follows.

(1) to (6) plus:

7. Thickness

    The layer thickness shall be in accordance with the manufacturer's instructions.

> Cementitious mortars and renders will not accommodate any movement of the structure. If movement is anticipated then this type of material is not appropriate.

### B6.4 SURFACE PROTECTIVE COATINGS

The coating system adopted should be suitable for application onto the existing reservoir substrate. Examples of the minimum performance requirements for the coating film are as follows.

1. Ability to span cracks

    The coating shall be able to span cracks of up to 0.3 mm in the substrate. In addition, the coating shall have the capacity to span unanticipated cracks with a cyclic movement of up to 0.2 mm that occur after application of the coating.

2. Adhesion

    Bond strength to the substrate, which has been prepared according to the manufacturer's instructions, shall be greater than 2.0 N/mm$^2$, when tested in accordance with ASTM D 4541-85. The coating shall be compatible for application to a damp substrate and shall have the capacity to remain bonded to the substrate under a hydrostatic head of 1 m on the concrete/brickwork side of the coating and a hydrostatic head equivalent to the full height of the walls plus 1 m applied directly to the coating.

> Elastomeric products will stretch and tear rather than give a true adhesion value.

3. Abrasion

    The wear index for the coating shall be greater than 100. The coating shall be tested in accordance with ASTM D4060-84. Taber abrasion resistance shall be determined with CS17 wheels, 1000 g load with 1000 cycles. The panel size to be used for testing shall be 100 mm square with rounded corners.

> Taber abrasion resistance values may be misleading on thermoplastic materials.

4. Impact

    The coating shall have an impact resistance of greater than 15 J when tested in accordance with BS3900: Part E7: 1974 (1986). The coating shall be tested 7 days after application or after the manufacturers recommended curing period, whichever is greater.

5. Permeability

   The $CO_2$ resistance of the coating shall be greater than a 50 m equivalent air layer thickness. The water vapour resistance shall be less than 20 g/m²/day.

   > $CO_2$ resistance will not be important where the concrete remains permanently saturated.

6. Chemical resistance

   The material shall be resistant to concentrations of chemicals considered appropriate to the specific project, during and possibly after construction, but which may include: dilute acids, dilute alkalis, soft water, sulphates and chlorides.

7. Water immersion

   The coating shall be suitable for permanent immersion in water.

8. Application temperature

   As considered appropriate for the specific project.

9. Potable water

   The coating (including primers, etc.) shall comply with Regulation 25 of the Water Supply (Water Quality) Regulations 1989, as amended (see also the note attached to Section B6.1 (10)).

## B6.5 METALWORK PAINT SYSTEMS

Examples of the minimum performance requirements for a metalwork paint system are as follows.

1. Impact resistance

   The paint shall demonstrate no deterioration when tested in accordance with BS3900: Part E3: 1973 (1991), using a 4.75 kg tup of diameter 14 mm through a drop of 570 mm. The paint shall be tested 7 days after application or after the manufacturers recommended curing period, whichever is the greater.

2. Hardness

   The surface hardness shall be greater than 100 when tested by BS3900: Part E9: 1973 (1991).

3. Water immersion

   The paint shall be suitable for permanent immersion in water.

4. Chemical resistance

   The material shall be resistant to concentrations of chemicals considered appropriate to the specific project, during and possibly after construction, but which may include: dilute acids, dilute alkalis, soft water, sulphates and chlorides.

5. Application temperature

   As considered appropriate for the specific project.

6. Potable water

   The paint (including primers, undercoats, etc.) shall comply with Regulation 25 of the Water Supply (Water Quality) Regulations 1989, as amended. (See also the note attached to B6.1 (10).)

## B6.6 CRACK INJECTION

The material used for crack injection (here termed grout) should be suitable for use with the crack widths to be filled. Examples of the minimum performance requirements for a crack injection grout are as follows.

1. Movement accommodation

   Where required the grout shall allow movement appropriate to the specific project.

2. Suitability for use in damp conditions

   The grout shall adhere to damp concrete with an adhesive strength greater than 3.0 N/mm$^2$. In addition, the grout shall be capable of displacing water in the crack, and of hardening in the presence of fresh water and groundwater.

3. Compressive strength

   The compressive strength of the grout at 28 days shall be equivalent to or greater than that of the parent concrete when tested in accordance with BS2782: Part 3: Method 345A: 1979.

4. Shrinkage

   The total shrinkage of the grout at any age up to a maximum of 6 months, tested at 20 °C and at 50% relative humidity, shall not be greater than 0.04%.

5. Tensile strength

   The tensile strength of the grout at 28 days shall be equivalent to or greater than that of the parent concrete when measured in accordance with BS6319: Part 7: 1985.

6. Chemical resistance

   The material shall be resistant to concentrations of chemicals considered appropriate to the specific project, during and possibly after construction, but which may include: dilute acids, dilute alkalis, soft water, sulphates and chlorides.

7. Viscosity

   To suit the specific application.

8. Chloride ion diffusion

   The coefficient of chloride ion diffusion resistance shall be less than $1 \times 10^{-12}$ m$^2$/s.

9. Application temperature

    As considered appropriate for the specific project.

10. Potable water

    The material (including primers, etc.) shall comply with Regulation 25 of the Water Supply (Water Quality) Regulations 1989, as amended (see also the note attached to Section B6.1 (10).)

## B6.7 CRACK SEALANTS

Examples of the minimum performance requirements for a crack sealant are as follows.

### B6.7.1 Flexible crack sealants

1. Adhesion to substrate

    If the concrete substrate cannot be guaranteed to be dry, then the sealant shall be suitable for application to damp concrete. Adhesion shall be adequate to withstand a hydrostatic pressure equivalent to the full height of the walls plus 1 m, with a minimum peel strength of 3.0 N/mm when tested in accordance with BS3712: Part 4: 1991.

2. Resistance to damage

    Once placed, the sealant shall be resistant to puncturing, creasing, de-bondment and any other damage from rubber-soled boots and light, rubber-tyred vehicles such as wheelbarrows.

3. Chemical resistance

    The material shall be resistant to concentrations of chemicals considered appropriate to the specific project, during and possibly after construction, but which may include: dilute acids, dilute alkalis, soft water, sulphates and chlorides.

4. Water immersion

    The sealant shall be suitable for permanent immersion in water.

5. Application temperature

    As considered appropriate for the specific project.

6. Extension

    Linear extension shall be greater than 25% when tested in accordance with ISO 10591: 1991.

7. Potable water

    The material (including primers, etc.) shall comply with Regulation 25 of the Water Supply (Water Quality) Regulations 1989, as amended (see also the note attached to Section B6.1 (10).)

---

Further guidance on the selection and specification of crack/joint sealants is given in CIRIA Technical Notes 128[16] and 144[17], CIRIA Special Publication 80[18], CIRIA Funders Report FR/CP/17[20], BS6213 and BS4254.

### B6.7.2 Non-flexible crack sealants

The performance requirements for a non-flexible joint sealant would be similar to those for crack injection (Section B6.6) without the requirements for compressive strength or viscosity.

## B6.8 JOINT SEALANTS

The performance requirements for a joint sealant would be similar to that for a flexible crack sealant given in Section B6.7.1.

## B6.9 LARGE VOLUME GROUTING

The performance requirements for large volume grouting would be similar to those for crack injection given in Section B6.6, but should not preclude the use of cementitious grouts.

# PART C

## Information to aid the decision making process

SECTION C1 CONTAMINATION OF RESERVOIRS
SECTION C2 INVESTIGATION AND TESTING METHODS
SECTION C3 CONTRACT DOCUMENTATION
SECTION C4 QUALITY MANAGEMENT IN REPAIRS TO SERVICE RESERVOIRS
SECTION C5 GUIDANCE FOR FUTURE DESIGN

# C1 Contamination of reservoirs

## C1.1 LEGAL REQUIREMENTS

In England and Wales drinking water standards are set by the Water Industry Act, 1991, which requires that licensed companies supply an adequate and continuous supply of 'wholesome' water. The general obligation to supply wholesome water is supplemented by the Water Quality Regulations, which specify the minimum standards against which the quality of water supplied for drinking, washing and cooking purposes is to be assessed, and incorporates the obligations on the UK in respect of European Community drinking water Directive 80/778/EEC.

The Water Quality Regulations impose requirements in respect of:

- microbial parameters including
    - faecal coliforms
    - total colony counts (at 22 °C and 37 °C)

- physico-chemical parameters including
    - pesticides
    - aluminium
    - lead
    - chlorine compounds (trihalomethanes (THM), etc.)
    - polyaromatic hydrocarbons (PAH)

- organoleptic parameters including
    - taste
    - odour

- undesirable parameters including
    - nitrates.

The 1989 Water Act (incorporated into the Water Industry Act 1991) created the criminal offence of providing water through pipes which is unfit for human consumption, with proceedings being brought by either the Secretary of State or the Director of Public Prosecutions. The Secretary of State is obliged to take enforcement action when standards are not met. However, he has the discretion not to enforce the regulations or to allow relaxations under certain restricted situations.

The fundamental aim of these regulations is to ensure that the water that is supplied to the consumer is fit to drink, thereby preventing the spread of water-borne diseases, and is free from contaminants.

## C1.2 WATER TREATMENT

In order that the requirements of the Water Industry Act, 1991, can be met it is generally necessary to treat the abstracted raw water to remove any biological and/or chemical contamination. A secondary reason for water treatment is to make it less corrosive to equipment and materials in the supply system. The extent of treatment required depends on the source (degree of contamination) of the water. Surface water generally requires more extensive treatment than groundwater, which has been naturally filtered by passage through the soil and rocks of the aquifer.

The main treatment processes include:

1. Storage in impounding reservoirs, where suspended materials can settle out and many types of bacteria die.

2. Screening to remove debris; some disinfection may also take place.

3. Aeration to remove odour and some dissolved gases, and to oxidise some dissolved metals for ease of filtration.

4. Coagulation and Flocculation to remove bicarbonates, trap bacteria and remove colour.

5. Sedimentation to remove the contaminants trapped by flocculation.

6. Filtration to remove particulate matter from the water and improve bacteriological quality.

7. Slow sand filtration: an alternative to (4), (5) and (6) above.

8. Granular activated carbon: added before (5) or (6) to remove tastes and odours by adsorption of chemicals.

9. pH adjustment to prevent corrosion of materials or deposition of scale.

10. Disinfection, usually by the addition of chlorine to remove harmful bacteria.

11. Softening: only carried out exceptionally by the exchange of magnesium and calcium salts for sodium salts.

12. Fluoridation: carried out in some areas to reduce the incidence of dental caries in accordance with the Water (Fluoridation) Act, 1985.

After treatment, the water that enters the supply system and is subsequently transported to service reservoirs should be free from either chemical or bacteriological contamination.

## C1.3 CONTAMINATION OF TREATED WATER

Contamination of the treated water can occur from two basic sources: the breakdown of materials used in the supply system, releasing toxic chemicals (the dwi has sponsored research into the leaching of metals from cement[15], which should be considered for both repair and new construction) or (more importantly) the ingress of untreated water into the system, carrying either toxic chemicals or bacteriological or viral organisms, which can multiply within the system. The growth of bacterial colonies may also lead to the premature breakdown (biodegradation) of the materials and components of the supply system, which in turn may release chemicals into the water supply.

The risk of contamination occurring in service reservoirs is probably greater than in many other elements of the water supply system. This is due to a number of features, including:

- The large surface area of the structure in contact with potentially contaminated groundwater.
- That there are entries for pipes and vents, etc., which pass through the structural materials and possibly form points of direct access.
- The large number of joints within the structure.
- The relatively large number of different types of material used within a reservoir, their susceptibility to application problems and long-term degradation.

Water quality analyses, primarily carried out by water quality scientists to ensure compliance with the statutory requirements, will detect contamination problems while the reservoir is still in service, and frequently instigate structural investigation and remedial works being carried out. Potential contamination problems that are developing but have not resulted in an unsatisfactory water quality analysis may be detected by the routine inspections carried out by most statutory undertakers (further details on the methods of investigation are given in C2). Structural problems may occur without a contamination problem, and these should not be overlooked.

Bacteriological or chemical contamination from groundwater ingress (predominantly through the roof) probably occurs most often, though accurate figures are not available. Contamination from the breakdown of materials used in construction is less frequent. All materials have a finite life, and the materials selected (see Section A4) should comply with the Water Quality Regulations to ensure that long-term exposure to the environment within the reservoir will not result in toxic substances being released or the material forming a nutritional medium or substrate for bacterial growth.

Achieving the required standards of water quality is not simply the function of the water quality chemist but is also the responsibility of the operations and maintenance staff, to ensure that contamination problems are minimised or eliminated completely.

### C1.3.1 Contamination from repair

The four independent certification/approval schemes in use for repair and waterproofing materials are described in Section A4.1. The way that these schemes should be taken into account, when selecting materials for use in contact with potable water, is briefly summarised below.

- Secretary of State's Approval (DoE Committee on Chemicals and Materials)

    The DoE-CCM advises the Secretaries of State on materials approval under Regulation 25 (1) (a) of the Water Act. The Committee will have assessed the potential for long- and short-term risk to the health of the consumer arising from the use of the product in contact with potable water. Approval may be subject to certain conditions, e.g. specified method of application or treatment after application.

- Water Byelaws Scheme

    Applies solely to the material's effect on the quality of potable water with which it may come into contact, as tested using short-term tests on material samples that have been applied and cured according to the manufacturer's recommendations. Materials having passed these tests are listed in the *Water Fittings and Materials Directory.*

- British Board of Agrément

    The BBA gives an independent performance-based assessment of the material's suitability for its intended purpose. The possibility of contact with potable water is not taken into account.

- Water Industry Certification Scheme

    This scheme certifies products manufactured in a plant operating a quality scheme in compliance with BS5750 (or equivalent) and the relevant water industry specifications.

Therefore, the fact that a material may be 'listed' in the *Water Fittings and Materials Directory* or have been approved by the Secretary of State should not be taken to mean that the separate components of the waterproofing or repair material systems in a 'non-hardened' state will not affect water quality. Indeed, contamination can be caused by minute quantities of chemicals such as pesticides and chlorinated phenols. It is considered that reservoirs should always be emptied if there is the likelihood of inward leakage of the components of repair materials. As well as the danger of contamination due to ingress of pollution and toxic vapours, any movement of materials and plant on the roof could give rise to dust and debris falling into the water. After completion of the work, reservoirs should be disinfected or flushed with water as appropriate[34]. A number of chemical cleaning agents are available, the use of which may be a useful alternative when high-pressure water cleaning is impracticable. If chemical cleaning agents are to be used, consideration must first be given to the effect of these agents on the receiving watercourse from the washout system, and any necessary precautions taken to protect it, e.g. dilution, pH correction, etc.

---

The following points should be taken into account, especially if emptying the reservoir is not operationally possible.

- Care must be taken during the application of sprayed or membrane waterproofing systems to avoid liquid components, cleaning fluids or solvents entering reservoirs through openings, joints, cracks or porous materials. Also, in hot weather apparently stable layers can melt.

- No mixing of repair materials, solvents, adhesives, primers, etc., must be carried out on the roofs of reservoirs. No containers should be stored or left on the roof where they may leak or be knocked over.

- Visual and, if necessary, chemical tests for leakage or spillage and water quality should be made, to check for the presence of uncured components of repair or waterproofing materials, which may enter reservoirs through cracked or porous concrete/brickwork, and at any joint in the walls and roofs or around hatchways, valve boxes or vent pipes.

- Care should be taken to avoid pollution of the external drainage system or natural watercourses during reservoir washout operations. The National Rivers Authority (NRA) should be consulted if there is any doubt as to water quality.

# C2 Investigation and testing methods

There are a number of levels at which an investigation of a structure should be carried out. The level of survey that is feasible is dependent on the circumstances at the time of the investigation: e.g. in the case of an emergency repair, where it is essential that the water supply should be maintained, time constraints may only allow a limited investigation.

Where a survey is to be carried out, it should be carefully planned and conducted in a systematic manner to determine the full extent and the cause or causes of any deterioration (see Figure C2.1). A full survey should be carried out in three stages to identify a **defect category**:

1. review of the existing plans of the structure
2. visual survey with limited testing
3. detailed investigation with a planned sampling and testing programme

However, it is seldom possible (and not always necessary) to undertake all three stages in a single investigation.

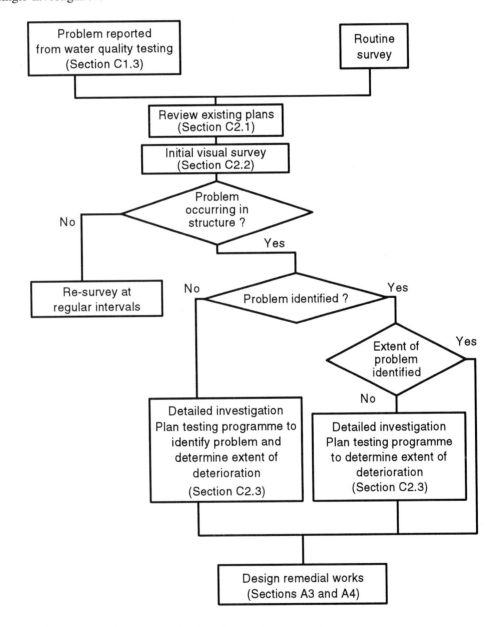

**Figure C2.1** *Stages in the investigation of a service reservoir*

Where possible, a routine survey should include stages (1) and (2).

Each of these stages is reviewed in the following sections. A number of publications have described the methods used in investigating a deteriorating concrete structure[35, 36]. It is not the intention of this document to reproduce these reports or describe the detailed methods used in an investigation.

## C2.1 REVIEW OF THE EXISTING PLANS

The existing plans of the structure should be reviewed, if they are available, for the following criteria.

- size, type (construction groups, see Section A1) and date of construction of the reservoir
- information on the materials used in construction
- unusual features of the design, which may affect the stability and durability
- determination of the number and position of joints
- orientation and size of the reinforcement and designed depth of cover for any future covermeter survey
- structure location and topography of the site
- the resources likely to be required to carry out the visual survey.

The reports of any previous investigations of either the structure or of the surrounding ground should also be reviewed prior to visiting the structure.

Structures are not always built as shown on the drawings, and for many, particularly older, structures plans are not always available.

The review may also be useful in preventing accidental overloading of the structure.

## C2.2 VISUAL SURVEY

The aims of the visual survey should be to identify the location and where possible the cause of any macro defects in the reservoir. The time period available for the inspection may be restricted to only one day from a typical seven day outage, which will, by necessity, restrict the amount of information that can be obtained. Experienced personnel should therefore carry out the investigation to maximise the quantity of information that can be collected. The reason for carrying out the investigation, whether it is a routine inspection or is being carried out in response to a failed water quality sample, will also influence the requirements of the survey. If the inspection is well planned, sufficient information can usually be gathered to prepare recommendations for any further work.

In order to carry out any level of investigation a certain amount of minimum basic equipment is necessary. This will include sufficient lighting to ensure that any defects can be detected, and where possible mobile staging to allow a close view of any suspect areas. The requirements for confined space working, in particular gas detection equipment, not working alone, etc., should be adhered to (see Section B1). A safe means of access for personnel and equipment should be provided. Access should not be blocked with equipment that may hinder a rapid exit in an emergency.

When undertaking an inspection, where time is at a premium, it is usually best to begin with an internal overview of the reservoir to identify any particular features that are likely to be susceptible to deterioration or identify any obvious problems, e.g. large-scale defects or staining. Where staging is available this can be positioned while the remainder of the structure is assessed.

A consistent referencing system should be established to accurately locate the position of any defects identified. This may be based on a grid system related to individual elements, where a relatively large number of small defects are present, or where only a few, more serious, defects occur may consist of simple measurements of the distance and direction from a prominent feature.

When defects are identified it is useful to categorise them in order of their seriousness and likely consequences for water quality and structural integrity. This will enable the detailed investigation or repairs to be focused on those features that are likely to require urgent work. Features which must be considered include the following.

*Internal*

- Cracking - either observed directly or indirectly from the products of water movement, e.g. staining, growth of stalactites, water dripping into the reservoir. The position, pattern, width, etc., should be noted.

- Spalling - fully or partially detached sections of concrete and any associated corrosion products.

- Corrosion (of all types of metalwork, including reinforcement) - in columns or beams where exposed, or staining of protective systems by corrosion products, ladders (particularly at welds), etc. The extent and loss of section should be recorded.

- Defective materials - in walls, columns or floors, e.g. softening of concrete or mortar detected by scraping the surface with a steel tool.

- Joint movement - opening or closing. The extent of movement should be recorded and condition of the sealant and filler checked.

- Joint sealants - the condition of any joint sealants should be established.

- Evidence of movement of structural elements - e.g movement of walls and floor slabs, often in association with joint movements.

- Overflow/washout system - checked for blockages.

*External*

- Perimeter drainage - checked for blockages, ponding on roof, etc.

- Condition of vents and access points - breakdown of mesh, vandalism, etc.

- Condition of the waterproof membrane - only if leakage is identified internally.

- Stability of earth mounds.

A limited amount of testing may be carried out as part of the inspection: e.g. a covermeter survey to give an indication of the depth of cover to steel, and carbonation testing on soffits.

The normal method of establishing the integrity of a waterproof membrane is to excavate trial pits above areas that are considered suspect. This is time-consuming, and may not reveal the location of the defect if water can travel beneath the membrane. Alternatively the integrity of any membrane may be established by hose tests to avoid the need to remove insulation. This must be planned well in advance in order that the necessary equipment can be available. There will also be a delay before any water will be detected entering the reservoir, and again the location of the defect will not be identified.

If the inspection was carried out as part of a routine series and no obvious defects were identified then no immediate action will be necessary. The reservoir should then be resurveyed at regular intervals as part of the normal schedule. If minor defects are observed, which do not threaten the integrity of either the structure or the water quality, then the merits of carrying out repairs immediately, delaying any work or resurveying at a shorter time interval must be considered. If serious defects are identified a detailed inspection will probably be necessary to establish the full extent of any work required.

## C2.3 DETAILED INVESTIGATION

When a defect has been observed in the structure, the cause and extent of the problem should be identified and remedial proposals developed. A detailed sampling and testing programme is usually required, which should be based on a grid system. This work could be carried out as part of a contract for remedial works or before a contract is let. It should be carried out either by experienced in-house staff or by specialist consultants, as some investigation techniques require specialist knowledge or experience[37, 38, 39] and due to the limited time period likely to be available for the investigation. As the extent of work tends to escalate when a thorough investigation is carried out, it is recommended that a full investigation be carried out prior to letting a contract for remedial works. This type of investigation would require that the reservoir or compartment be out of service for a period of 3 to 10 days (excluding drawdown and refilling) depending on the size and the intensity of testing.

### C2.3.1 Non-destructive tests

*(a) Delaminations*
Delaminations in concrete are detected by striking the surface and noting the change in sound emitted. The velocity of sound waves is different in air and concrete, resulting in a 'hollow' sound in delaminated concrete rather than the 'ringing' sound of intact concrete. With practice, it is possible to accurately identify the boundaries of the affected area, which should then be marked on the surface of the structure and on an appropriate grid sheet. Electronic methods, such as ultrasonic pulse velocity (UPV), may also be used. These can be used to detect discontinuities in the fabric of the element at greater depth than audible methods.

*(b) Carbonation*
The reduction in pH of cementitious systems is used to detect the occurrence of carbonation (see Appendix I Section AP I.2.2 (b)). A pH sensitive indicator (phenolphthalein) will change colour from colourless to pink over the pH range 8.4 to 10. This substance, when sprayed onto a freshly broken surface of concrete or mortar, turns pink responding to the alkalinity of the concrete. No change in colour indicates carbonated material. The depth of carbonation should be measured to the nearest millimetre and the results recorded.

> Care should be taken when using phenolphthalein in a 'live' reservoir, owing to its toxicity.

The holes that are produced in obtaining dust samples for chemical testing (see Section C2.3.2(a)) can be utilised for measuring carbonation depth, by knocking off the shoulders of the holes with a hammer and chisel (see Figure C2.2). Carbonation should be measured at four locations around the circumference of the hole. The drilled edges of these holes should not be used directly, as they may be contaminated with uncarbonated material[40] from the interior of the concrete and thus give an erroneous reading.

The results should be tabulated and used to plot a carbonation histogram.

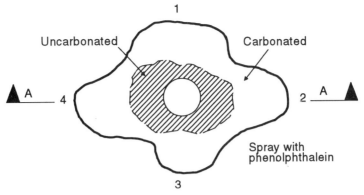

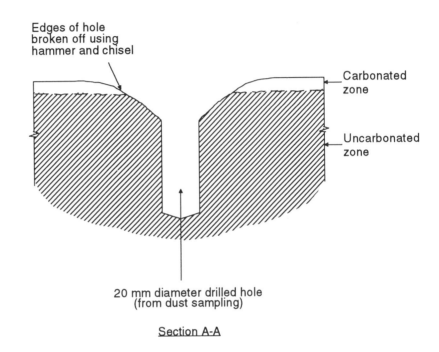

**Figure C2.2** *Measurement of depth of carbonation of concrete*

*(c) Concrete cover*
The depth of cover to the reinforcement and the orientation of the reinforcement (if this is not already known) can be determined by using an electromagnetic covermeter in accordance with the procedures laid out in BS1881: Part 204: 1988, provided that the diameter of the bars and the material type are known. The instrument measures the disturbance in a magnetic field, induced by the presence of a reinforcing bar. The magnitude of the effect is proportional to the size of the bar and its distance from the probe. Some modern equipment is capable of measuring the size of reinforcement if the depth of cover is known.

The depth of cover at the grid positions should be recorded and the data tabulated to produce a cover histogram. Special attention should be given to zones where the depth of cover is seen to be less than the depth of carbonation (see Section C2.3.1(b)).

Care needs to be taken with the interpretation of results, especially in areas of congested reinforcement.

*(d) Corrosion potential*

The corrosion of steel is an electrochemical process and thus the electrical potential of steel reinforcement will vary depending on the service environment of the metal. Under favourable conditions (of moisture content of the concrete, depth of carbonation, etc.) it is possible to measure the variation in electrical potential of steel reinforcement in concrete from one part of the structure to another. To make electrode potential measurements requires a suitable millivoltmeter and a reference electrode. The millivoltmeter is connected to the reinforcement and to the half-cell probe, which is placed in contact with the concrete surface. A complete circuit then exists and the difference in potential between that generated by the half cell and that of the reinforcement can be measured.

Results are usually reported in the form of half-cell potential contour maps, in which areas of high corrosion risk can be delineated[41]. Areas can be identified where corrosion is most probable; however, this should not be taken to imply that corrosion is occurring or that the corrosion rate is great. Results can be misleading in saturated concrete owing to oxygen starvation. Decisions on remedial works should not be based on the results of half-cell potential surveys in isolation; chloride, cover and carbonation results at the same locations must all be considered.

*(e) Crack movement*

Cracks caused by thermal effects or structural loading may respond quite quickly to changes in the structural environment. It is important to know if cracks are actively opening or closing (live) before repair or waterproofing is carried out. Cracks are likely to open when the reservoir is full and close when it is drawn down. This must be considered when taking measurements.

Various forms of electrical strain gauge are available, some of which are suitable for site use. However, the most practical method of detecting and measuring short-term crack movement is with a demountable mechanical strain gauge (demec gauge), which can be held on gauge studs bonded to the concrete surface, in accordance with BS1881: Part 206: 1986. Vibrating wire strain gauges have been found to be useful for observations over long periods of time.

Readings are quick and relatively easy to take and a series of measurements made over a 24 hour period is usually sufficient to establish whether a crack is live due to thermal effects.

*(f) Estimating the strength of in-situ concrete*

The strength of *in-situ* concrete may be estimated by a number of methods. It is beyond the scope of this document to describe in detail the types of test which may be used. However, they include the following.

*Rebound hammer* – This measures the rebound energy following a collision with the concrete surface, in accordance with BS1881: Part 202: 1986. The rebound energy can be correlated with the compressive strength of the concrete by testing on a large number of specimens, which encompass the range of strengths expected in the structure.

*Internal fracture test* – This measures the tensile force required for a wedge anchor bolt to cause failure of the concrete. Correlation with the compressive strength must be derived experimentally.

*Pull-out test* – This measures the maximum tensile force that can be sustained by an embedded insert. The pull-out force can be empirically related to the compressive strength as determined by standard tests in accordance with BS1881: Parts 116 and 120: 1983, but the correlation between the two must be established experimentally for the apparatus concerned in accordance with BS1881: Part 207: 1992.

*Pull-off test* – This measures the maximum force required to pull a metal block together with a layer of concrete or mortar from the surface of the concrete to which it is attached. This may be carried out in two ways[42]: by attaching the block directly to the surface, or by partially coring the concrete and attaching the block to the cylinder produced. Correlation with compressive strength must be derived experimentally for the type of concrete under investigation in accordance with BS1881: Part 207: 1992.

*Penetration resistance test* – This measures the depth of penetration of a metal probe fired into the surface of the concrete. Calibration must be undertaken for each concrete to be tested in accordance with BS1881: Part 207: 1992.

*Break-off test* – This measures a flexural tensile strength in a plane parallel to the concrete surface at a predetermined distance below the surface. Correlation with compressive strength must be established experimentally in accordance with BS1881: Part 207: 1992.

> All these tests are sensitive to the local conditions of the concrete (e.g. moisture content, aggregate type, etc.) and should not be considered as a substitute for direct measurements from concrete samples.

*(g) Leakage detection*
Outward leakage from a reservoir can usually be estimated by carrying out a drop test. The location of the individual leaks are not as easy to detect. In many cases the visual survey is the only means of identifying possible leaks. Acoustic and current flow surveys are increasingly being used to detect outward leakage while the reservoir is still partially full.

Acoustic surveys measure the noise generated by water egressing the reservoir, while the flow survey measures the current associated with water movement out of the reservoir. By taking measurements at a number of locations it is possible to locate isolated areas with high rates of leakage.

*(h) Void detection*
Voids beneath slabs or behind walls can be detected by ground-penetrating radar surveys or transient dynamic response (TDR). Both techniques will only detect relatively large voids (at present), and the results require skilled interpretation.

### C2.3.2 Types of sample
Sampling should be systematic (random samples give little information) with sample locations selected with respect to the results of the visual survey, and should be representative of the extent of deterioration observed. A number of types of sampling and testing are likely to be required to assess the condition fully.

*(a) Dust samples*
These may be obtained by drilling into the surface of the structure using a hammer drill fitted with a 20 mm diameter drill bit[43]. The dust should be collected at regular depth increments, usually 25 mm, and stored in labelled bags. These samples are used for chemical tests, in particular determination of the chloride and sulphate content of the material. The type and content of cement in concrete or mortar may also be analysed (see Section 2.3.3 (c)).

*(b) Core samples*
Cores should be obtained in positions indicated by an engineer. They can be used to obtain a measure of the quality of the materials in the roof or upper sections of the walls. The normal dimensions of cores are 100 mm diameter and at least 150 mm in length. The core should be washed, labelled and carefully logged before being packed for transportation.

The cores should be fully representative of the conditions of the structure, including areas that are thought to be sound. The cores are primarily used for strength testing, though they can also be used for petrographic examination (see Section C2.3.3 (e)).

Where cores are taken through the roof, the resulting core hole should be plugged with a suitable repair material, whose physical characteristics match those of the existing roof. In taking cores, cutting through the reinforcement should be avoided if possible.

*(c) Sawn samples*
Sawn samples are particularly useful for investigating the condition of any waterproofing, or to investigate the condition of the roof material at any cracks.

*(d) Exposure of reinforcement*
Concrete may be removed to expose the reinforcement, at selected locations, to estimate the extent of corrosion and measure the loss of cross-section of the reinforcement.

## C2.3.3 Laboratory tests

There are a number of tests that can only be carried out in the laboratory on samples that are obtained during the detailed investigation. These include the following.

*(a) Chloride content determination*
The chloride ion content of the concrete or mortar can be determined from dust samples by means of an acid dissolution followed by titration with silver nitrate solution, in accordance with BS1881: Part 124: 1988. The end point of the reaction is determined either chemically or electrically. The results of the analyses are used to produce a chloride profile into the material and thus to assess whether chloride contamination is likely to be the cause of the observed deterioration.

*(b) Sulphate content determination*
The sulphate content of any cementitious system can be determined from dust samples by either titremetric or gravimetric techniques, in accordance with BS1881: Part 124: 1988. Again, an indication of whether sulphate attack has caused the deterioration can be obtained.

*(c) Determination of cement type*
If a low cement content is suspected and if the aggregate type is known, it may be appropriate to determine the cement type, and approximate content, in the original mix based on the chemical data, in accordance with BS1881: Part 124: 1988. If, based on results of strength tests, it is thought that high alumina cement (HAC) was used and there is the likelihood of concrete strength reduction due to the conversion of the hydrated cement to a weaker form, then the degree of conversion can be determined by differential thermal analysis (DTA).

*(d) Strength tests*
Crushing tests should be carried out in accordance with BS1881: Part 120: 1983 on cores taken from deteriorating and sound areas of the structure, thought to represent the range of compressive strengths expected in the roof. Where the concrete is apparently of uniform quality, a smaller number of samples may be tested.

The compressive strength results should be compared with the strength specified on the original drawings where possible. Wide variations in strength indicate local areas of deterioration (or conversion in the case of HAC). Values of less than 20 $N/mm^2$ represent poor-quality concrete. Concrete that contains horizontal cracks in the upper parts of the core (e.g. from freeze-thaw action; see Appendix I Section AP I.3) may show a high compressive strength but still be of poor quality.

*(e) Petrographic examination*
Petrographic examination of either concrete or brickwork samples can give diagnostic information on the causes of a number of types of deterioration. It is not possible to give a full description of the methods employed in this document. The types of deterioration that can be identified are:

- AAR
- sulphate attack
- salt crystallisation damage of concrete, mortar or bricks
- frost damage
- aggregate shrinkage
- porous concrete or mortar
- microcracked and delaminated concrete or mortar

Information can also be obtained on the mix proportions of the concrete or mortar.

> All testing should be undertaken in an approved laboratory, which should, where possible, operate an accepted quality assurance scheme, e.g. BS5750 (ISO9001), or be NAMAS accredited.

# C3 Contract documentation

It is good practice to use standard conditions of contract to ensure that all contractual points are covered in a way familiar to the tenderer, while enabling the specifier to concentrate on the points specific to the contract. However, work of a specialist nature is not generally covered by preprinted standard contract documentation. This therefore means that a large number of special specification clauses are often necessary.

## C3.1 PREPRINTED DOCUMENTATION

There are a number of standard conditions of contract and specifications available, which can be used as the basis for a contract into which special conditions could be added.

### C3.1.1 Standard conditions of contract

*(a) Institution of Civil Engineers (ICE)*
The ICE produce two standard conditions of contract:

- ICE Conditions of Contract and Forms of Tender, Agreement and Bond for use in conjunction with Works of Civil Engineering Construction, Sixth Edition, January 1991

- ICE Conditions of Contract, Agreement and Contract Schedule for use in connection with Minor Works of Civil Engineering Construction, First Edition, January 1988.

*(b) Joint Contracts Tribunal for the Standard Form of Building Contract (JCT)*
There are a number of variants of the JCT Standard Form of Building Contract (local authority, private usage, etc.): each consists of three variants plus subcontract documents. The 1980 Edition has had numerous amendments made to it (1984 to 1991). This form of contract is only appropriate where an architect or contract administrator is employed.

*(c) General Conditions of Contract for Building and Civil Engineering (GC/Works)*
The third edition of the Government Conditions of Contract for Building and Civil Engineering Works (GC/Works/1) came into use in 1990, and is intended for major works of new construction, whether of building or civil engineering. There are many different forms of the contract.

### C3.1.2 Standard specifications

*(a) Water Authorities Association*
The Civil Engineering Specification for the Water Industry, Fourth Edition (CESWI) was published in 1989 by the Water Research Centre plc on behalf of the Water Authorities Association. CESWI is regarded as the standard document for civil engineering contracts let by Water Undertakers, although it is not universally adopted for reservoir repair and waterproofing.

It is intended for use with the ICE Conditions of Contract and the Civil Engineering Standard Method of Measurement.

*(b) Department of Transport (DoT)*
Probably the most complete UK contract documentation is that produced by the Department of Transport (DoT) and published by Her Majesty's Stationery Office. The most recent edition, *Manual of Contract Documents for Highway Works*, was issued in December 1991 and comprises the following volumes.

| | | |
|---|---|---|
| Volume 0*: | Model Contract Document for Major Works and Implementation Requirements | |
| | Section 1: | Model Contract Document for Highway Works |
| | Section 2: | Implementing Standards |
| | Section 3: | Advice Notes |
| | | |
| Volume 1: | Specification for Highway Works | |
| | | |
| Volume 2: | Notes for Guidance on the Specification for Highway Works | |
| | | |
| Volume 3: | Highway Construction Details | |
| | Section 1: | Carriageway and Other Details |
| | Section 2: | Safety Fences and Other Barriers |
| | | |
| Volume 4: | Bills of Quantities for Highway Works | |
| | Section 1: | Method of Measurement for Highway Works |
| | Section 2: | Notes for Guidance on the Method of Measurement for Highway Works |
| | Section 3: | Library of Standard Item Descriptions for Highway Works |
| | | |
| Volume 5*: | Contract Documents for Specialist Activities | |
| | Section 1: | Model contract Document, Standard, Specification, Notes for Guidance and Bills of Quantities for Topographical Surveys (not issued) |
| | Section 2: | Maintenance Painting of Steel Highway Structures |
| | | |
| Volume 6: | Departmental Advice Notes on Contract Documentation and Site Supervision | |
| | Section 1: | Construction (Design and Management) Regulations 1994 Requirements for Health and safety Plan |
| | Section 2: | Use of Substances Hazardous to Health in Highway Construction |
| | Section 3: | (not issued) |
| | Section 4: | Amendments. |

* Published by the DoT on behalf of the Overseeing Departments.

*(c) National Building Specification (NBS)*
The Code of Procedure for Project Specification (a code of procedures for building works) sets out basic principles including:

- the nature and purpose of specification
- specification by performance or prescription
- specification of options and alternatives
- specification by reference or description
- materials, products and purpose made components
- workmanship
- arrangement
- the use of language.

The code also contains procedural recommendations on producing the specification and coordination of the specification with drawings, bills of quantities and schedules.

There are three versions of the specification: standard, intermediate and minor works. Each comprises a number of volumes/sections, which are updated separately on a regular basis. This specification is only appropriate for building works.

*(d) Property Services Agency (PSA) General Specification*

The PSA General Specification, Third Edition, 1991, is a library of specification clauses for use when producing project specifications for:

- projects estimated to cost more than £150,000

- specialist Building and Civil Engineering including 'soft' landscape work estimated to cost more than £25,000 and £70,000 respectively

- projects estimated to cost less than £150,000, where the technical content of individual work sections of the PSA Minor Works Specifications is considered inadequate for the particular circumstances.

There is also a PSA Minor Works Specification for Building and Civil Engineering Work, Fourth Edition, 1992. This specification is only appropriate for building works.

*(e) Concrete Society Technical Report No.38*

Concrete Society Technical Report No.38, Patch Repair of Reinforced Concrete Subject to Reinforcement Corrosion, 1991, contains a model specification for concrete repair work.

### C3.1.3 Methods of measurement

*(a) Department of Transport (DoT)*
A method of measurement for highway works is included as a part of Volume A of the *Manual of Contract Documents for Highway Works*; refer to Section C3.1.2. (b).

*(b) Institution of Civil Engineers (ICE)*
The ICE Civil Engineering Standard Method of Measurement (CESMM), Third Edition, provides a library of standard bill items for civil engineering work.

*(c) Concrete Society Technical Report No.38*
Concrete Society Technical Report No.38, Patch Repair of Reinforced Concrete Subject to Reinforcement Corrosion, 1991, contains a method of measurement for concrete repair work.

*(d) Concrete Repair Association*
Concrete Repair Association Method of Measurement for Concrete Repair, 1990, provides a library of items for use in bills of quantities for repair work, including: surface cleaning, surveying, patch repairs, crack repairs, pore/blowhole fillers, levelling mortars/fairing coats, coatings and resin injection.

## C3.2 SUITABILITY OF DOCUMENTATION FOR WATERPROOFING AND REPAIRING SERVICE RESERVOIRS

None of the above sets of documents is ideal for service reservoir waterproofing and repair work without modifications. A system is needed that requires as few alterations as possible and does not involve the tenderer or specifier buying volumes of expensive, mostly inappropriate, clauses. The industry would benefit by adapting the most readily applicable existing documents to provide a model contract document. The following are suggested as suitable for most reservoir repair contracts.

### C3.2.1 Conditions of contract

- ICE Conditions of Contract and Forms of Tender, Agreement and Bond for use in conjunction with Works of Civil Engineering Construction, Sixth Edition, January 1991

or

- ICE Conditions of Contract, Agreement and Contract Schedule for use in connection with Minor Works of Civil Engineering Construction, First Edition, January 1988.

### C3.2.2 Specification

- WAA, Civil Engineering Specification for the Water Industry, Third Edition, 1989.

with special clauses for repair work taken from:

- Concrete Society Technical Report No.38, Patch Repair of Reinforced Concrete Subject to Reinforcement Corrosion, 1991.

### C3.2.3 Method of measurement

- ICE Civil Engineering Standard Method of Measurement, Third Edition.

with use, for repair work, of:

- Concrete Repair Association Method of Measurement for Concrete Repair, 1990.

## C3.3 CONTRACT FORMAT

Recommendations for standard contract format should include:

- instructions to tenderers
- document cover
- index and table of contents
- form of tender
- conditions of contract
- special conditions
- specification
- bill of quantities.

# C4 Quality management in repairs to service reservoirs

## C4.1 INTRODUCTION

The incorporation of formal quality systems into the management of waterproofing and repair activities provides a framework for effective, demonstrable, project management.

Advantages are as follows:

- Project planning is aided from inception through to implementation and monitoring systems are established.

- Coordination between project stages is defined, together with the project structure and the responsibilities of the participants.

- Documentation is generated systematically.

- All interactions of parties are considered and suitable arrangements made.

- There is a systematic framework for communicating and agreeing the means by which requirements are to be met.

- Monitoring of performance is developed and documented, and used to improve procedures.

- Implementation and continuing effectiveness of the quality system are monitored through quality audits.

Quality audits may, and almost certainly will, be carried out internally (first-party audits). In addition, in order to provide confidence to other parties, audits may be carried out by the client (second-party audits), or by an organisation independent of the contract (third-party audits). The scope and extent of the audit programme will vary according to the sensitivity of the project and the need to verify compliance.

Contractors/suppliers can seek independent certification for their in-house systems from any of a number of third-party certification bodies. This has the advantage of minimising duplication of auditing effort and providing immediate recognition that a quality system has been developed and implemented. It is, however, necessary to confirm that the certification applies to the particular material, process or activity under consideration.

## C4.2 DURING THE INVESTIGATION AND CONDITION ASSESSMENT

The implementation of a quality system during this phase has the effect of promoting a planned and systematic approach, which would lead to greater effectiveness and therefore economy.

The key elements of the system would be:

- definition of areas requiring investigation/survey
- identification of qualified staff on the basis of aptitude
- defining procedures relevant to the investigation
- issuing procedures and project-specific instructions to staff
- carrying out the investigation and reporting in a standard format
- collating records for use in the next stage.

## C4.3 DURING DESIGN OF THE REPAIR

During the design phase, the use of a formal quality system would require that information, incorporated into the analysis, is assessed for status (e.g. preliminary, unchecked, verified, unadjusted, adjusted).

The key elements would be:

- preparation of the 'repair intent' and its review against the requirements

- coordination of the summary of the design predictions, the construction specification and the original performance specification, including the internal and external environmental conditions.

The simplified flow diagrams in Figures C4.1 and C4.2 show design stage activities carried out concurrently with material selection and testing activities.

The contract documentation should include a requirement that sufficient details of the tenderer's quality assurance system be submitted to ensure its suitability for the contract.

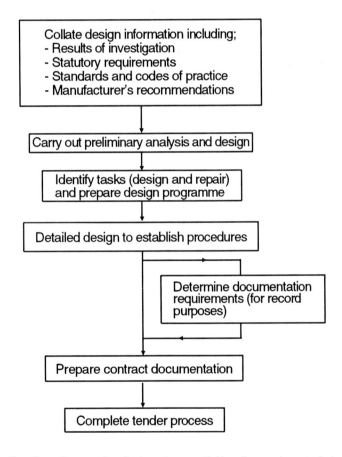

**Figure C4.1** *Outline flow diagram for design-stage activities (for each part-site)*

## C4.4 DURING MATERIALS SELECTION AND TESTING

Quality management is particularly suited to the selection and testing of materials. This covers the activities of both the Water Undertaker and the materials supplier.

A requirement for documentation to substantiate the conformance of materials to specification can be included in the supply contract. It could also be stipulated that the supplier operates within a formal quality system, thus providing confidence of conformance to the specification.

Such a system would include documented materials control and traceability procedures. It would also identify the inspection and test regime and the procedures to be followed in the event that non-conforming materials were identified.

The adequacy of the quality system could be substantiated through independent quality audits (a requirement for access to carry out audits should be included in the contract documentation) or through reliance on third-party certification by a recognised body.

As most repair materials are susceptible to deterioration, the quality system should include procedures for the handling, storage, packaging and delivery of the material to the repair site.

Where relevant material performance data are not available, a suitable testing regime must be devised. Under these circumstances, the testing can be considered to be governed by the activities flow chart indicated in Figure C4.2.

Materials testing would be required at various stages in the process of identification, specification and implementation of the repair programme. These stages would require definition in terms of specific testing requirements and the records of tests carried out. The materials supplier's system should also encompass the verification that testing equipment is suitably calibrated.

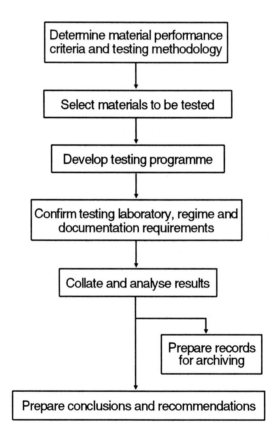

**Figure C4.2** *Outline flow diagram for materials selection and testing activities*

## C4.5 DURING THE REPAIR

The requirements of quality management during the repair phase can be considered to consist of two distinct systems, that of the contractor and that of the employer. The contractor's system should include procedures for the management and implementation of the work, to provide a service that conforms to the contractual and specification requirements.

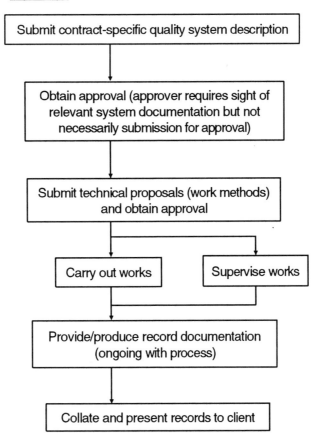

**Figure C4.3** *Outline flow diagram for installation-stage activities - Contractor*

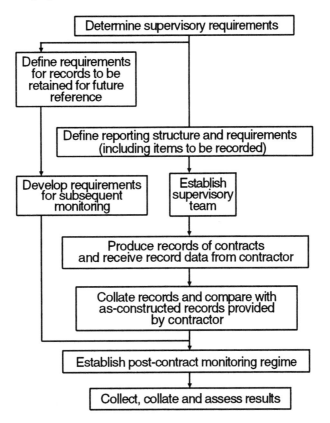

**Figure C4.4** *Outline flow diagram for installation-stage activities - Employer*

The key elements would be:

- creation and review of method statements
- verification of the training of operatives for the use of particular materials, repair systems and techniques
- verification and testing of material properties
- supervision of activities
- full records of the measurement of parameters required to meet objective criteria (including storage and post-placement requirements).

Method statements should pay particular attention to matters such as:

- drainage
- surface preparation
- jointing
- material properties and quality design detail prior to placing and curing
- matters relating to water and vapour protection to all metalwork in construction
- manufacturer's instructions and the avoidance of damage
- careful execution of the repair works
- careful stripping of formwork, especially during the plastic stage
- provision made for adequate relative movement between the roof and walls.

The employer's system would include procedures required to supervise the work and to provide records of the repairs carried out and monitoring procedures. The employer's system may also include subsequent monitoring procedures to assess the effectiveness and durability of the repairs. Flow charts are given in Figures C4.3 and C4.4 relating to the repair stage activities for the contractor and the employer respectively.

## C4.6 QUALITY SCHEME FOR REPAIR CONTRACTORS

The Concrete Repair Association (CRA) has produced for its members guidelines on BS5750 (ISO 9001) *Quality Systems*. This includes a format for *Quality Assurance and Quality Procedures Manuals*. They also have a proforma for members to report periodically on their quality system development and implementation. There has also been a recommendation to the members of the CRA that they should adopt a sector scheme approach in developing a QA initiative within its organisation. It should also be noted that the CRA has issued a Code of Practice for members.

# C5 Guidance for future design

(It is assumed that reinforced or prestressed concrete construction will be the future norm for the construction of reservoirs.)

Reservoirs are generally regarded as long-life structures, and therefore design for serviceability and durability is the main concern of BS8007[44]. The recommendations for designers given below are stimulated by the content of the previous sections in this Manual and therefore cover details specific to durability. Many of the recommendations will require modifications to be made to other aspects of the reservoir, and hence careful consideration of all these features should be given during the design phase.

## C5.1 DRAINAGE

1. All reservoir roofs should have a minimum fall of 1 in 100 (preferably 1 in 80) and be covered with a waterproofing membrane (see Section C5.7 (1)). The membrane may be of the bonded, unbonded, hydrophilic or liquid applied types and should overlap the roof/wall top joint by at least 300 mm. The type of membrane should be selected to suit the particular project.

> Unbonded membranes are susceptible to damage in windy conditions, until fully ballasted, and sometimes from thermal effects. Active membranes may require modifications to the make-up of the insulation layer.

2. Roof water should flow to an effective perimeter drainage system, which should be well clear of the walls, to carry the water away from the structure. Water flows down the outside of walls to a low-level drain should be avoided.

3. Any 'back-of-wall' (e.g. fin drains) or underfloor drainage systems must be separate from the roof drainage system.

## C5.2 PROTECTION OF STRUCTURAL METALWORK

1. Once a reservoir has been in service (and sometimes prior to commissioning at certain times of year), a damp internal environment is the norm. It is therefore important that new construction should include a long-life protective system to all metalwork susceptible to corrosion, particularly mild steel. The metalwork should be detailed so that ponding of water and trapping of condensation is avoided.

2. Where possible, metalwork should be treated with factory-applied coatings, as a higher level of quality control can be expected than from site-applied coatings.

3. Combinations of different types of metal that result in galvanic action should be avoided, or suitable electrical insulation measures taken.

## C5.3 AVOIDANCE OF CONCRETE DETERIORATION

1. Concrete materials, mix designs and cover requirements should be selected for durability, with particular regard to the likely aggressive agents present in the environments (internal and external) for the particular reservoir: e.g. pH (of the stored water), carbonate imbalance, carbonation, sulphate attack, (see Appendix I Section AP I.2 for details).

2. The filling of shutter-tie holes must be carefully executed. An appropriate material should be selected that will achieve adequate compaction through the full wall thickness and which will not shrink away from the edges of the hole. It must also be ensured that thorough compaction of concrete is obtained around shutter ties.

3. Similarly, it is important to control the quality of design detail, materials and compaction when grouting-up around pipes, ducts, etc., that pass through the structure and have been boxed out for later grouting.

## C5.4 AVOIDANCE OF CRACKING (OTHER THAN FORMED JOINTS)

1. Proper consideration of the factors likely to induce cracking in a reservoir should be given during the design stage: e.g. possible roof uses, potential for subsidence.

2. Detailed measures to control cracking are contained in BS8007, and a summary of the factors that help to prevent or control early-age cracking are given in Table 10 of CIRIA Report No. 91[45].

3. If it is necessary to cast in lifts, each lift should be placed as soon after the previous lifts as possible and at intervals of less than 3 days, before the concrete in the first lift has cooled and shrunk significantly or becomes too dry.

4. The concrete should be specified for the lowest possible heat of hydration. The addition of pulverised fuel ash (pfa) or ground granulated blast-furnace slag (ggbs) would increase setting times and control the heat of hydration, while observing the minimum cement content requirements.

> The rate of gain of strength would be slower for concrete produced using these types of cement, particularly during cold weather.
>
> Some continental ggbs has been reported to contain high concentrations of heavy metals, which may be leached by some types of water and would make them unsuitable for use in a service reservoir[15].

5. Water reducing admixtures may improve the resistance of the concrete to cracking by reducing the water:cement ratio and hence shrinkage.

## C5.5 AVOIDANCE OF CORROSION OF REINFORCEMENT

1. Attention is drawn to BS8007: 1987, Clauses 2.7.4 (durability), 2.7.5 (impermeability) and 2.7.6 (cover). It is recommended that the depth of cover to the reinforcement be checked during construction.

2. The protection afforded by the specified cover and a correctly designed and adequately compacted concrete mix is satisfactory for the majority of construction. Increases in cement content, increases in cover or the use of special reinforcement have design implications, which must be checked.

## C5.6 AVOIDANCE OF JOINT FAILURE

1. Roof/wall joints (particularly sliding joints) should be carefully designed to prevent leakage and root penetration.

2. The recommendations for joint types and spacings given in BS8007 are a major part of the design philosophy and should be carefully considered in relation to the design approach adopted.

## C5.7 DURABILITY OF WATERPROOFING MEMBRANES

1. The provision of a waterproofing membrane over all reservoir roofs is strongly recommended. It is justified by the long-life requirement of this type of structure, the possibility of cracking at a later date and the high cost of application at a later stage. However, such a covering must not be a substitute for a properly designed roof complying with BS8007 to retain aqueous liquids, or an excuse for poor workmanship in constructing the roof. If in doubt, prior to applying the membrane, the roof should be tested for watertightness.

2. Membranes should always be installed in accordance with the manufacturer's instructions. They should be suitably detailed to avoid damage over movement joints, at corners or by covering with unsuitable insulation material.

3. Only proven materials should be used on large roofs. New materials should be restricted to small areas of application for trial purposes.

4. Bonded membranes should be left loose over contraction joints, and a similar provision should be made for liquid membranes.

## C5.8 ACCESS (INCLUDING ACCESS LADDERS), VENTILATION, OPENINGS AND ANCILLARY ITEMS

1. Stairway access is always preferable to ladder access and should be provided wherever possible.

2. The security of the reservoir must be carefully considered in all aspects of design, but in particular when designing access openings and ventilation provisions. The use of access covers that incorporate ventilation openings is recommended.

3. There have been instances reported where ventilated covers have later been replaced with plain covers, as an oversight; it is important to ensure that the dual purpose of the access cover is recorded.

4. Access openings should be designed such that rescue equipment can be satisfactorily erected above an access opening (or a special opening should be provided) to facilitate the rescue of injured personnel.

5. Careful consideration should be given to sealing waterproofing membranes around upstands and roof vents.

6. Wherever possible, Grade 316S31 stainless steel, complying with BS970: Part 1: 1983 or BS1449: Part 2: 1983, should be used within reservoirs, particularly for ladders or protection around deep openings at inspection water levels.

7. The number of ventilators provided should be sufficient to prevent unacceptable pressure changes from occurring during routine operation of the reservoir. Guidance on the number of ventilators required is given in Section A3.7.1. At least one ventilator should be provided in any area that can be isolated within the reservoir by rising water levels.

8. Ensure that the positioning of pipes and sealing details does not impose restraint, unless it has been considered during the design.

## C5.9 WATERSTOPS

1. Where waterstops are included within a reservoir they should be water tested prior to the reservoir being commissioned. The testing should be applied directly to the waterstop: i.e. before any joint sealant has been installed.

> Testing in this manner, may increase construction time and costs by making sealant application more difficult, as the joint will require thorough drying before the sealant is installed, unless a damp-tolerant sealant system is used.
>
> BS8007: 1987 states: *'it is not necessary to incorporate waterstops in properly constructed construction joints'*. It is, however, considered a prudent measure to include a waterstop in all types of joint.

## C5.10 DURABILITY OF SEALANTS

1. Materials should be selected that do not support significant bacterial growth and which comply with the water quality regulations (see Section A4.1).

2. The selection of sealants for particular requirements of flexibility and movement should be based on the total movement figure quoted by manufacturers.

3. The procedures used to sterilise reservoirs during their return to service should be designed to prevent concentrated sterilising solution from coming into direct contact with sealants (further advice on sealant selection and usage is given in Section A4.2.8).

# References

1. JOHNSON, R.A, LEEK, D.S AND KING, E.S.
   *Waterproofing and repairing underground reservoir roofs: A report of current practice with recommendations*
   CIRIA Technical Note TN145, London, 1991

2. WHITE, E.
   *Survey of concrete potable water storage structures*
   WRc Report ER302E, 1988

3. HMSO
   Reservoirs Act, 1975
   HMSO, London

4. PAINT RESEARCH ASSOCIATION
   *Quality control procedures when blast cleaning steel*
   Paint Research Association, November 1980

5. BUILDING RESEARCH ESTABLISHMENT
   *Repairing brick and block masonry*
   BRE Digest 359, 1991

6. HEYMAN, J
   *The masonry arch*
   Ellis Horwood Ltd, Chichester, 1982

7. MOTT MACDONALD
   *CTAP manual for the assessment of masonry arch bridges*
   Mott MacDonald, 1990

8. HARVEY, W.J
   Application of the mechanism analysis to masonry arches
   *The Structural Engineer*, Vol. 66, No. 5/1, 1988

9. CRISFIELD, M.A AND PACKMAN, A.J
   *A mechanism program for computing the strength of masonry arches*
   Transport and Road Research Laboratory, Research Report 124, Crowthorne, 1987

10. VASSIE, P.R.
    The influence of steel condition on the effectiveness of repairs to reinforced concrete
    *Construction and Building Materials*, Vol. 3, No. 4, 1989

11. WATER AUTHORITIES ASSOCIATION
    *Civil Engineering Specification for the Water Industry, 3rd Edition*
    Water Authorities Association, 1989

12. HMSO
    Water Industry Act, 1991 (consolidating the Water Act 1989)
    HMSO, 1991

13. HMSO
    The Water Supply (Water Quality) (Amendment) Regulations, 1991
    HMSO, 1991

14. BRITISH STANDARDS INSTITUTION
    *Suitability of non-metallic products for use in contact with water intended for human consumption*
    BS6920, 1988

15. LAWRENCE, C.D.
    *International review of the composition of cement pastes, mortars, concretes and aggregates likely to be used in water retaining structures*
    HMSO, 1994

16. CONSTRUCTION INDUSTRY RESEARCH AND INFORMATION ASSOCIATION
    *Civil engineering sealants in wet conditions*
    CIRIA, Technical Note TN128, London, 1987

17. AUBREY, D.W.
    *Performance of sealant-concrete joints in wet conditions - Results of a laboratory testing programme. Volume 1: Main results & discussion*
    CIRIA, Technical Note TN144, London, 1992.

18. CONSTRUCTION INDUSTRY RESEARCH AND INFORMATION ASSOCIATION
    *Manual of good practice in sealant application*
    CIRIA/BASA (joint publication), Special Publication SP80, London, 1992

19. AUBREY, D.W.
    *Performance of sealants in wet conditions - joint surfaces*
    CIRIA, Report R128, London, 1992

20. CONSTRUCTION INDUSTRY RESEARCH AND INFORMATION ASSOCIATION
    *Design and construction of joints in concrete structures*
    CIRIA, Funders Report FR/CP/17, 1994

21. SKIPP, B.O AND HALL, M.J.
    *Health and safety aspects of ground treatment materials*
    CIRIA, Report R95, London, 1982

22. HEALTH AND SAFETY COMMISSION
    Control of Substances Hazardous to Health (COSHH) Regulations
    Health and Safety Commission, 1988

23. JACKSON, P.
    Leaks in concrete structures at formwork ties
    *Concrete*, November 1966

24. BUILDING RESEARCH ESTABLISHMENT
    *Sulphate and acid resistance of concrete in the ground*
    BRE, Digest 363, p8, 1991

25. AMERICAN CONCRETE INSTITUTE
    *Cement and concrete terminology, manual of concrete practice, Part 1: Materials and general properties of concrete*
    ACI, Committee 116, 1987

26. PARROTT, L.J.
    *A review of carbonation in reinforced concrete*
    C&CA, for BRE, DoE, Building Research Establishment, 1987

27. CONCRETE SOCIETY
   *Alkali Silica Reaction - Minimising the risk of damage to concrete. Guidance notes and model specification clauses*
   Concrete Society, Technical Report No.30, (Ed M.R. Hawkins), p34, 1987

28. CONCRETE SOCIETY
   *Non-structural cracks in concrete*
   Concrete Society, Technical Report No. 22, (Ed. C.D. Turton), p39, 1982

29. CONSTRUCTION INDUSTRY RESEARCH AND INFORMATION ASSOCIATION
   *Concrete in the Oceans - Coordinating report on the whole programme*
   CIRIA (Ed. M. Leeming), 1989

30. BUILDING RESEARCH ESTABLISHMENT
   *The durability of steel in concrete: Part 1. Mechanism of corrosion*
   BRE, Digest 263, p8, 1982

31. THE BUILDING ADVISORY SERVICE (for the Building Employers Confederation)
   *Confined Spaces*
   Building Advisory Service, Construction Safety, Section 23

32. BRITISH STANDARDS INSTITUTION
   *Workmanship on building sites, Part 14: Code of practice for below ground drainage*
   BS8000, 1989

33. BRITISH STANDARDS INSTITUTION
   *Guide to the durability of buildings and building elements, products and components*
   BS7543, 1992

34. WATER AUTHORITIES ASSOCIATION
   *Operational guidelines for the protection of drinking water supplies*
   Water Authorities Association

35. PULLAR-STRECKER, P.
   *Corrosion damaged concrete - assessment and repair*
   CIRIA, Butterworths, 1987

36. AMERICAN CONCRETE INSTITUTE
   *Guide for making a condition survey of concrete in service*
   ACI, Committee 201-1R-68, 1984

37. BUNGEY, J.H.
   *Testing concrete in structures - A guide to equipment for testing concrete in structures*
   CIRIA, Technical Note TN143, London, 1992

38. LEEMING, M.B.
   *Standard tests for repair materials and coatings for concrete - Part 2: permeability tests*
   CIRIA, Technical Note TN140, London, 1993

39. CONSTRUCTION INDUSTRY RESEARCH AND INFORMATION ASSOCIATION
   *Standard tests for repair materials and coatings concrete - Part 3: stability, substrate composition and shrinkage tests*
   CIRIA, Technical Note TN141, London, 1993

40. ROBERTS, M.H.
   *Carbonation of concrete made with dense natural aggregates*
   BRE, Information Paper IP6/81, 1981

41. AMERICAN SOCIETY FOR TESTING AND MATERIALS
    *Half-cell potential of reinforcing steel in concrete*
    ASTM, C876-80, 1980

42. McLEISH, A.
    *Standard tests for repair materials and coatings for concrete - Part 1: pull-off tests*
    CIRIA, Technical Note TN139, London, 1993

43. ROBERTS, M.H.
    *Determination of chloride and cement contents of hardened concrete*
    BRE, Information Paper IP21/86, 1986

44. BRITISH STANDARDS INSTITUTION
    *Design of concrete structures for retaining aqueous liquids*
    BS8007, 1987

45. HARRISON, T.A.
    *Early-age thermal crack control in concrete (Revised edition)*
    CIRIA, Report R91, London, 1992

# Index

| | Page |
|---|---|
| Access, ventilation, openings and ancillary items | |
| - design guidance | A25, A28, A64, C25 |
| - during inspection | B1, B3, C5, C7 |
| - problems, causes and effects | A14, A25, B5 |
| - repair method and material performance | A25, A52, A53 |
| - safety aspects | B1, B3, C2, C6 |
| Acid attack | A12, A43, A55, A57, B19 |
| Active membranes | see Hydrophilic membranes |
| Aggregate unsoundness | A57, C13 |
| Alkali aggregate reaction (AAR) | A55, C13 |
| Aluminium | 1, A52, A53, A54, C1 |
| Approval schemes | A15, A30, A41, A51, B23, C3 |
| - British Board of Agrément | A30, A32, C3 |
| - Secretary of State | A30, A51, B25, C1, C3 |
| - Water byelaws scheme | A30, A31, B25, C3 |
| - Water Industry Certification scheme | A30, A32, C3 |
| Arch roofs | 1, A18, B10 |
| - assessment | A18 |
| - load relief | A18, B10 |
| | |
| Backfilling | A38, B11 |
| Bonded membranes | A4, A22, A24, A32, A35, A36, A37, A38, A62, A63, B12, B13, B14, B22, B25, B26, C23, C25 |
| Bonding coats | A33, A34 |
| Box-outs | A54, C24 |
| Brickwork | |
| - chemical attack | A11, A13, A17, A54, A55 |
| - construction groups | A1, A2, A4, A5, A6, A11, A12, A17 |
| - defect categories | A5, A10, A11, A12, A15, A17, A19, A43, A54, A59, A60, B8, C4 |
| - deterioration; causes and effects | A10, A11, A17, A54, A59, C13 |
| - physical attack | A11, A17, A57 |
| - repair methods and materials | A17, A43, B8, B24, B27, B30 |
| - roofs, assessment of | A18 |
| - stress state change of | A18, A60, B9, B10 |
| | |
| Carbonate imbalance | A12, A57, A58, C23 |
| Carbonation | A12, A21, A43, A55, C7, C8, C9, C10, C23 |
| Cast iron | A2, A4, A6, A11, A52, A54 |
| Chemical attack | A11, A12, A13, A17, A21, A54, A55, A62 |
| Chloride attack | A12, A21, A34, A53, A54, A56, A57, A60, A62, B17, C10, C11, C12 |
| Coating, protective | A6, A17, A21, A32, A33, A43, A44, A51, A55, B15, B20, B22, B29, B30, B31, C16, C23 |
| | |
| Columns | |
| - construction groups | A2, A4, A5, A6, A11, A12, A13 |
| - defect categories | A11, A12, A13, A22, A62, C7 |
| - investigation | A22, C5, C7 |
| - methods of repair | A16, A17, A21, A22 |
| - repair activity, problems | B8, B10 |
| - repair materials | A33, A43, A44, A45, A46, B19, B23 |

CIRIA Report 138

Concrete
- beam and reinforced slab     A2, A4, A5, A11, A12, B7
- chemical attack     A11, A12, A13, A17, A21, A54, A55, A62
- cover     A21, A62, C6, C7, C9, C10, C23, C24, C25
- cracking     A7, A8, A10, A12, A13, A14, A15, A18, A19, A20, A24, A25, A32, A35, A36, A42, A56, A57, A58, A59, A62, C,7, C10, C24
- design guidance     A1, A2, A3, A4, A5, A6, A7, A22, A35, A62, C23, C24, C25
- detailed investigation     A9, A45, A46, A60, C5, C8, C18
- deterioration; causes and effects     A11, A12, A14, A17, A21, A54, A59, A62, C23
- drying shrinkage     A12, A60
- *in-situ* reinforced flat slab     A2, A4, A5, A12
- mass concrete     A1, A2, A4, A5, A6, A11, A12, A18, A21, A60, B8
- plastic settlement     A12, A60, A62
- plastic shrinkage     A60
- precast     A2, A5, A6, A12, A20
- prestressed     A2, A5, A6, A12, A21, C23
- repair methods     A15, A17, A18, A19, A21, A22, A25, A30, A33, A34, B1, B11, B12, B19, B21, C14, C18
- repair performance     A32, A51, B19, B23, C3, C18
- repair systems and problems     A17, A21, A23, A25, A32, A33, B12, B20, B22, B24, C22
- roofs, assessment of     A18
- spalling     A8, A11, A12, A21, A62, C7
- visual survey of     A9, A20, C5, C6, C11

Construction groups     A1, A4, A5, A6, A11, A12, A13, A14, C6

Contamination
- of reservoirs     A8, A9, A12, A22, A25, A56, B5, B6, C1
- of treated water     A8, A11, A12, A14, A19, A28, A31, A44, A45, CA, C2
- through repair activity     B11, C3

Contract documentation     A30, C14, C19, C20
Contract procedures     B21, C14

Corrosion
- bimetallic     A11, A16, A26, A52, A53
- of reinforcement     A8, A10, A11, A12, A15, A19, A21, A55, A56, A60, A61, A62, C7, C10, C12, C24
- of structural metalwork     A10, A11, A14, A15, A16, A26, A52, C2, C7, C23

Corrosion potential     C10
Cost estimates     B19, B21, B22
Cover     see Concrete - cover

| | |
|---|---|
| Cracking | |
| - causes and effects of | A7, A8, A12, A13, A14, A15, A18, A19, A25, A36, A45, A46, A56, A57, A58, A59, A60, A61, A62, A63, B5, B8, B9, B10, B12, B13, B15, C4, C7, C10, C12, C13, C24 |
| - design guidance | C18, C19, C23, C24 |
| - effect on leakage | A8, A12, A19, A56, A60, A61, B8 |
| - effect on reinforcement | A8, A12, A19, A62 |
| - effect on waterproofing | A12, A19, A37, A40, A63, B10 |
| - effect on water quality | A56, B8, B11, C1 |
| - fillers and sealants | A19, A38, A42, A46, A47, B15, B20, B22, B33, B34 |
| - injection of | A19, A24, A45, A46, B32 |
| - investigation | A10, A19, A59, C7, C10, C12, C13 |
| - repair methods, materials and performance | A19, A25, A35, A36, A37, A39, A40, A42, A43, A45, A46, A47, B13, B26, B28, B29, B30, B32, B33, C16 |
| Curing | |
| - of concrete | A58, A63, B12, B24, C22 |
| - of liquid membranes | A41, B29, C22 |
| - of metalwork paint systems | A16, B31, C22 |
| - of sealants | A47, A48, C22 |
| - of surface protective coatings | A42, B30, C22 |
| Defect categories; causes and effects | 4, 12, 13, 16, A10, A15, C5 |
| Delaminations | A17, A55, A61, C8, C13 |
| Department of Transport Specification | C15, C16 |
| Detailed investigations | B20, C5, C7, C8 |
| Disinfection | B11, C2, C4 |
| Documentation, owners | A30, C14, C15, C16, C17, C18, C19, C20 |
| Drainage | |
| - floors | A8, B5, B7 |
| - general | A24, A35, A40, A43, A63, B5, B11, B17, B18, B28, C1, C7, C22, C23 |
| - roofs | A4, A24, A35, A37, B12 |
| - underfloor | A8, B7 |
| - walls | A6, A8, A46, B6, B7 |
| Drying shrinkage | A12, A60 |
| Durability, maintenance and cost | A9, B19, B20, B21 |
| Earthworks | B17, B18 |
| Efflorescence | A59 |
| Environment | |
| - internal service | A11, A45, A47, A52, A53, A62, B2, C3, C19, C23 |
| - external service | A32, A53, B17, C19, C23 |
| Floors | |
| - assessment of | A8, B7, C6, C7 |
| - construction groups | A2, A6, A7 |
| - defect categories | A11, A12, A13, A14 |
| - leakage through | A8 |
| - methods of repair | A19, A22, A32, B7 |
| - repair activity, problems | A15, A42, A48, A60, A61, B8 |
| - repair materials | A35, A42, B25, B29 |
| Ferrous metals | A52 |
| Fillets | A37, A38 |
| Freeze-thaw | A12, A61, C12 |

| | |
|---|---|
| Gravel | A21, A40, A54, A56, B17, B28 |
| Glass reinforced plastics | A1, A37 |
| Ground granulated blastfurnace slag (ggbs) | A45, C24 |
| Ground settlement | A12, A14, A60, A61 |
| Grout | |
| - cement based grouts | A46, A47, A50 |
| - chemical grouts | A46, A50, A51 |
| - design guidance | A23, A46, A54, B16, C24 |
| - large volume | A19, A23, A46, A50, B22, B34 |
| Health and Safety | A1, A31, A50, B1, B2, B3, C3, C15 |
| Hydrophilic membranes | A32, A36, A37, A40, A42, A63, B25, B27, C23 |
| Inspections | see Visual Survey |
| Instrumentation | C9, C10 |
| Joint(s) | |
| - failure of | A8, A10, A12, A13, A14, A18, A22 |
| - fillers and sealants | A22, A33, A48, B6, B33, B34 |
| - material durability and unit costs | B20, B22 |
| - problems of filling and sealing | A13, A28, B11, B15 |
| - repair methods and procedures | A22, A23, A24, C22, C25 |
| Laboratory tests | |
| - investigation | A8, A22, A24, B3, B20, C5, C7, C8, C10, C11, C12, C13 |
| - materials | A16, A31, A32, A38, A39, B12, B19, B23, B24, B25, B26, B27, B28, B29, B30, B31, B32, B33, B34, C3, C19, C20, C22 |
| - water quality | A8, A9, B11, C1 |
| Leaching | A42, A43, A50, A51, A55, A58, B5, C2, C24 |
| Ladders | A14, A25, A26, A28, A52, A53, C7, C25, (see also Access) |
| Leakage detection | |
| - in membranes | A38, A39, A40, C4 |
| - in reservoirs | A8, A9, A10, B7, C11 |
| Lighting, temporary | C6 |
| Liquid membranes | A19, A22, A35, A36, A37, A38, A41, A63, B13, B22, B28, B29, C4, C23, C25 |
| Loading | A12, A13, A18, A19, A28, A35, B11, B13, B18, C6, C10 |
| Low compressive strength concrete | A56 |
| Maintenance of repairs | B19, B20, B21 |
| Material categories | A15, A30, A32, B20, B22 |
| Material selection procedure | A15, A30, A33, A46, A51, B19 |

| | |
|---|---|
| Membranes | see Hydrophilic membranes, Bonded membranes, Liquid membranes or Unbonded membranes. |
| Metalwork | |
| - aluminium | see Aluminium |
| - corrosion of | see Corrosion |
| - design guidance | A16, A26, A52, C23 |
| - detailed investigation | C5, C8, C9, C10, C18 |
| - ferrous | A52 |
| - galvanised | A52, A53 |
| - paint systems and performance | A16, A26, A44, B15, B20, B22, B31 |
| - reinforcement corrosion | A8, A10, A11, A12, A19, A21, A55, A56, A60, A61, A62, C7, C10, C12, C16, C24 |
| - repair materials, unit costs | B22 |
| - repair methods | A16, A21 |
| - visual survey | C6 |
| National Building Specification (NBS) | C15 |
| Non-destructive tests | A38, C8 |
| Overflow | A9, A28, C7 |
| Paint systems | see Metalwork - paint systems and performance |
| Permit to work | B3 |
| Petrographic examination | C11, C13 |
| Physical attack | A11, A17, A54, A57 |
| Pipework | A9, A14, A28, A60, B5 |
| Plans, existing | C5, C6 |
| Plastic settlement | A12, A60, A62 |
| Plastic shrinkage | A60 |
| Pollution | |
| - of external drainage | A50, C4 |
| - of water supply | see Contamination |
| Pressure grouting | see Grout |
| Property Services Agency General Specification | C16 |
| Pulverised Fuel Ash (pfa) | A45, C24 |
| Quality management | |
| - during design of the repair | C19 |
| - during materials selection and testing | A32, C19, C20 |
| - during repair | C20 |
| - investigation and condition assessment | C18 |
| - schemes for repair contractors | C20, C21, C22 |
| Reactive aggregate | A12, A56, A60 (see also Alkali aggregate reaction (AAR)) |

Reinforced concrete
- columns — A2, A6, A11, A12, A22, A62, C7
- corrosion of reinforcement — A8, A10, A11, A12, A15, A19, A21, A55, A56, A60, A61, A62, C7, C10, C12, C24
- design guidance — A5, A7, B1, B2, C19, C20, C21, C23, C24, C25, C26
- deterioration, causes and effects — A10, A11, A17, A21, A54, A60, C12, C13, C23
- durability estimates — B19, B20, C23, C24
- floors — A2, A7, A12, A21, A22, A60, A62, C7
- investigation and testing — A21, A45, A60, C3, C5, C6, C7, C8, C9, C10, C11, C12, C13
- repair methods and materials — A21, A30, A32, A33, A34, A35, A51, B12, B19, B20, B21, B22, B23, B25, C3, C16, C17, C18, C19, C20, C22
- repair system components — A33, A34, A35
- retaining walls — A2, A5, A6, A8, A11, A12, A54, A56, A57, A59, A60, A62
- roofs — A2, A4, A5, A8, A19, A21, A40, A60

Reinforcement corrosion, design guidance — C24

Roof
- assessment of — A17, A18, A19, A21, A28, A35, A42, A55, A60, A61, A62, B5, B10, B11, B18, C12
- construction activity — A4, A5, A24, A37, A40, C22
- construction group and material — A1, A2, A3, A4, A5
- defect categories — A11, A12, A13, A14, A28
- drainage provisions — A24, A35, A40, B5, B6, B12, C7, C23
- insulation and earthworks — A22, A37, B10, B17, B18
- Internal waterproofing treatment — A15, A19, A21, A22, A42
- leakage — A8, A38, A40, A59, B5, B11, C3, C4
- repair activity, problems — A22, A40, A41, A63, B8, B9, B10, B11, B12
- repair methods and materials — A15, A22, A37, A38, A39, A40, A42, A43, A63, B20, B21, B22, C25
- insulation removal — A15, B10
- surface coatings — A4, A5, A36, A37, A38, A39, A40, A41, A42, A43, A44, B28, B29

Roots, plant — A13, A41, A60, A61, A63, C25

Safety
- hazards — A11, A26, A50, A51, B1
- of operations — A25, A31, B1, B2, B3, B4

Salt crystallisation — A12, A57, A61, C13

Sealants
- durability — A13, A22, A23, A39, A47, A48, B6, B15, B20, B22, C26, see also Cracking, Joints
- materials types and specification — A19, A22, A23, A32, A40, A46, A47, A48, B33, B34
- non flexible — A19, A32, A46

Screeds
- problems with — A41, B12
- types of — A35, A36, B22

Seeded earth — B17
Stairs — see Ladders (see also Access)
Sterilisation — see Disinfection
Strength of *in-situ* concrete — A56, A60, C10, C11, C12
Surface coatings — A17, A21, A42, A43, A44, B15, B20
Sulphate attack — A12, A55, A60, C12, C13, C23

| | |
|---|---|
| Surveys | see Visual survey and Detailed investigation |
| Temperature effects | A22, A46, A48, A56, A57, A62, B9, B13, B14, B19 |
| Testing | see Detailed investigation, Laboratory tests, Non-destructive tests, Test samples |
| Test samples | C11, C12 |
| Thermal movement | A12, A13, A22, A45, A46, A60, A62, B13, C10, C23 |
| Unbonded membranes | A37, A39, A40, A41, A63, B12, B13, B27, C23 |
| Upstand | A14, A28, A37, A39, C25 |
| Ventilation, requirements | A14, A25, A28, A29, A45, A64, A65, A66, A67, B1, B3, C26 |
| Void detection | C11 |
| Visual survey | B20, C4, C5, C6, C7 |
| Walls | |
| - construction group | A2, A5, A11, A12, A13, A14 |
| - crack sealing | A19 |
| - defect categories | A11, A12, A13, A14 |
| - drainage provisions | B6, B18, C23 |
| - earthworks | B7, B11, B18 |
| - investigation and testing methods | C4, C7, C11 |
| - joints; roof/wall | A8, A18, A22, A28, A39, B9, B10, C25 |
| - leakage through | A8 |
| - repair activity, problems | A15, B8, B10, B11 |
| - repair materials | A18, A22, A33, A35, A37, B24, B25, B26, B27, B28, B29, B30, B31, B32, B33, B34 |
| - repair methods | A15, A22, C22 |
| - thermal movement of | A22, A62 |
| - waterproofing treatments | A5, A35, A37, A38, A39, A40, A41, A42, A63, B25, B26, B27, B28, B29, B30, B31, B32, B33, B34 |
| Washout | C4, C7, (see also Disinfection) |
| Water Authorities Association | C14 |
| Waterbars | see Waterstops |
| Waterproofing systems | |
| - breakdown; causes and effects | A13, A62, A63, C7 |
| - contract for repair; suggested documents | C14, C16 |
| - cost estimates of repair work | B19, B20, B21, B22 |
| - damage of | A13, A37, A38, A39, A41, A63, B10, B11, B14, B15, B17, C23, C25 |
| - membranes; types of | A4, A5, A6, A36, A37 |
| - performance of | A30, A35, A36, A37, A38, A39, A40, A41, A42, B25, B26, B27, B28, B29, B30, C25 |
| - problems of | B10, B11, B12, B14, B15, C4, C23 |
| - quality management | C18 |
| - repair methods and materials | A17, A19, A22, A24, A35, A36, A37, A38, A39, A40, A41, A42, A57, B15 |
| Waterstops | A13, A23, A49, A50, A54, B16, C26 |
| Water quality requirements | A8, A9, A30, A31, C1 |
| Water treatment | C1, C2 |
| Workmanship | A11, A13, A41, A54, A63, C15, C25 |

CIRIA Report 138

CIRIA Report 138